FARM TRACTORS IN COLOR

FARM TRACTORS
In Color

Michael Williams

drawings by
John W. Wood
Brian Hiley
W. Hobson

MACMILLAN PUBLISHING CO., INC.
New York

To David

Library of Congress Cataloging in Publication Data

Williams, Michael, 1935–
Farm tractors in color.

(Macmillan color series)
Bibliography: p.
1. Tractors—History. 2. Tractors—Pictorial works.
3. Agricultural machinery—History. I. Title.
S711.W5 1975   631.3'72'09   74-20544
ISBN 0-02-629300-5

Macmillan Publishing Co., Inc.
866 Third Avenue, New York, N.Y. 10022
First American Edition 1974

Text printed and book bound
by Richard Clay (The Chaucer Press) Ltd
Bungay, Suffolk, England
Color illustrations produced by Colour Reproductions, Billericay.

CONTENTS

ACKNOWLEDGEMENTS

This book has been produced with the help and co-operation of many people. I would particularly like to thank the following for their assistance: The Royal Agricultural Society of England for permission to quote from their *Journal*; Allis-Chalmers, Milwaukee, Wisconsin, for providing original material on the development of tractor tires; Mr. Frank Smith of Frampton, Lincolnshire, for his permission and assistance in photographing tractors from his collection, plates 25, 26, 38, 48, 49, 50, 53, 54, 55, 57, 59, 61, 62, 67, 71, 84; The *Farmers Weekly* for plates 22, 27, 35, 36, 39, 51, 52, 56, 58 showing tractors owned by Mr. John Moffitt, Stocksfield, Northumberland; Deere & Co., Moline, Illinois, for plates 57, 60, 78, 79, 88, 89; International Harvester Co., London, for plates 17, 18, 20, 103; Klockner-Humboldt-Deutz AG, Cologne, for plates 40, 77, 113, 114, 115, 116; The Smithsonian Institution, Washington, D.C., for plates 21 and 98; the Collections of Greenfield Village and the Henry Ford Museum for plates 7, 37, 68, 69; the Director, The Science Museum, London, for plate 9; Mr. T. J. Fowler, Brandon, Manitoba, for plates 6 and 19.

Illustrations were also supplied by the following companies: Aebi, British Leyland, County, Fendt, Fiat, Howard Rotavator, Lely, Massey-Ferguson, Mercedes Benz, New Idea, Valmet and Volvo.

Tractors in the collection of Mr. Frank Smith were photographed by Peter Adams, as were plates 63, 64, 65, 66, 76, 81, 82, 83, 85, 86, 87, 90, 91, 93, 94.

M. W.

I

THE IRON HORSE

The efficiency of agriculture affects all of us. When cultivation was merely stirring the soil with a pointed stick, food production was so slow and laborious that almost everybody had to work in the fields to produce enough for survival. Animal power raised the work rate appreciably and released substantial numbers of people for other jobs, such as building railways or waging war. In 1890, when steam power was just beginning to make a slight impact on agriculture and when five or six of the first tractors ever built were working in South Dakota, 42·6 per cent of America's labour force was engaged in agriculture, working with a vast force of mules and horses.

The development and increasing use of farm tractors has helped to accelerate the release of manpower from the land. For developing countries mechanisation, together with other methods of improving food production, offers the hope of better living standards for millions of people.

Steam power might have achieved the same result as the internal combustion engine, but the rate of progress would have been slower. The lumbering steam traction and ploughing engines of seventy years ago were limited by their weight and their appetite for fuel and water. The development of lighter, more efficient steam engines with high-pressure tubular boilers, arrived too late to affect agriculture. Tractors by then were firmly established—light, versatile, reasonably reliable and cheap enough for many farmers to buy.

The first tractors were made in America, but the development of tractors has been an international affair. The Otto engine, with its four-stroke cycle, came from Europe and was the basis of tractor development. The features which make a modern tractor efficient and versatile, such as the power-take-off, hydraulic lift, crawler tracks or rubber tyres, have come from many countries. The development is continuing, but meanwhile we owe much to the men who have already contributed to the history of the tractors which help to produce our food.

America

Credit for producing the first practical tractor is generally given to the Charter Gas Engine Company of Chicago, Illinois. Their first machine consisted of one of their own single-cylinder gasoline engines, designed for stationary work, which was mounted on the chassis and wheels of a Rumely steam engine. The tractor was equipped with a reverse gear and with two massive flywheels to overcome the irregular running of the engine. The first tractor was completed in 1889 and transported to a ranch in South Dakota for testing. In spite of the limitations of exposed gears with primitive lubrication, the machine appears to have been a success, and five or six similar tractors were produced by the Charter Company for farmers in the same area.

As gasoline engines became more widely accepted in America, and as steam power attracted increasing demand, interest in tractor development grew and in 1892 several experimental machines were produced. The Case Threshing Machine Company of Racine, Wisconsin, one of the leading American steam traction engine manufacturers, produced a single tractor in 1892, which was equipped with a twin-cylinder horizontal engine. The machine, which was designed by William Patterson, never got beyond the experimental stage. The make-and-break ignition remained unreliable, and the Case Company lost interest and concentrated on their steam engines which were outstandingly successful.

Also in 1892, John Froelich, an Iowa contractor, designed a gasoline tractor which was to have a far-reaching influence. Froelich was born in 1849 in Girard, Iowa, but set up business in the small, neighbouring community of Froelich, which was named after his father. John Froelich earned his living from hiring out equipment on a contract basis. He had well-digging equipment, operated a grain elevator and also worked with his steam traction engine and threshing tackle. It appears that his threshing operations extended into South Dakota, where he may have seen or heard of the Charter Company tractors working.

His first tractor was powered by a single-cylinder, vertical engine with bore and stroke of 14 inches which was made by Van Duzen of Cincinnati. It was mounted on a Robinson chassis, and was equipped with gearing for one forward and one reverse speed. The tractor was taken to South Dakota in its first season, and completed a threshing run of fifty days totalling 72,000 bushels of wheat.

Backed by this success, John Froelich got support for a company to manufacture and market tractors. The headquarters was at Waterloo, Iowa, and the company was called the Waterloo Gasoline Traction Engine Company. Tractor manufacturing was not the success that John Froelich had hoped, and the company had to turn to stationary engines in order to stay in business. Although Froelich withdrew from the company, it continued to experiment with tractors from time to time. In 1913 they produced the successful LA model, followed by the Model R, which was the forerunner of the popular Waterloo Boy. John Deere purchased the Waterloo Company in 1918 as a means of entering the tractor market.

The Van Duzen Company, which had provided the engine for the original Froelich tractor, produced their own prototype with a single-cylinder engine in 1894. A peculiarity of this tractor was that it had a whistle. Perhaps this allowed the driver to signal to threshing machine crews, or helped former steam engine drivers feel at home. Van Duzen was taken over by the Huber Manufacturing Company of Marion, Ohio, a well established steam engine company. Huber combined a modified Van Duzen engine and their own well-tried traction engine transmission in 1898, to launch themselves on the tractor market. The result, helped no doubt by the Huber name, was a commercial success and thirty sales are recorded for their first year. Huber remained in the tractor business for more than forty years.

Another nineteenth-century development with far-reaching effects was S. S. Morton's experimental tractor of 1899. Morton used a single-cylinder engine mounted horizontally on a specially designed chassis. This was possibly the first break from the established principle of using trac-

9

tion engine components. Morton's ideas eventually attracted the attention of the Ohio Manufacturing Company of Upper Sandusky, Ohio. In 1905 the Ohio Company signed an agreement to supply tractors to the recently formed International Harvester Company. This marked International Harvester's entry into the tractor market, although one of the companies incorporated into International, Deering Harvester Company, had experience of self-propelled equipment, including mowers and an experimental corn picker, dating back to 1891.

Several other individuals and companies produced experimental tractors before 1900. A few, such as the Hocket, Otto and Flour City, achieved a modest volume of sales during this period.

There is some controversy over which American company was the founder of the world tractor industry. Most of the earliest machines were either experimental and based on traction engine components, or were produced in small numbers. On the evidence available, the Huber Company appears to have the strongest claim, although the Ivel Company in Britain may have been the first to establish production on a full commercial basis. Most authorities, however, award the honour to the Hart-Parr Company.

Charles W. Hart and Charles H. Parr were both engineering students at the University of Wisconsin where they started experimental work with gasoline engine design. They formed a company to manufacture the oil-cooled valve-in-head stationary engine they had developed. This company was later reorganised and moved to Charles City, Iowa, where the first Hart-Parr tractor was produced in 1901. The engine for this tractor was oil-cooled, and had two cylinders with 9-inch bore and 13-inch stroke. Hart-Parr No. 2 followed in 1902, again with a twin-cylinder engine, operating at the slow speed of the period—280 r.p.m. Tractor No. 3 followed in 1903, and this machine is preserved in the Smithsonian Institution, Washington D.C. It was the forerunner of the 17–30-h.p. model produced in the same year, of which fifteen were sold. The Hart-Parr Company produced a remarkable series of tractor models before becoming part of the Oliver Farm Equipment Company in 1929. With few exceptions, these tractors were heavily constructed, some weighing as much as twenty-six tons, and they achieved a reputation for reliability and durability in the prairie lands for which they were primarily designed.

During the first decade of this century, American tractor production, from scores of manufacturers, steadily gained momentum as reliability and performance improved. The industry was greatly helped by the fact that American farmers accepted the advantages of tractor power far more readily than their European counterparts.

An outstanding name in the development of tractors of this period was C. M. Eason. He produced prototype tractors from 1905, but was not involved on a commercial scale until 1912 when he was a designer with the Wallis Tractor Company of Cleveland, Ohio. Eason is given much of the credit for the design of the Cub tractor, the first model produced by Wallis in 1913. The outstanding feature of the Wallis Cub was the introduction of 'frameless' construction. A curved sheet of boiler plate steel formed the crankcase and transmission housing. This served both as protection for the working parts and as the main strength or backbone of the tractor. The frameless design, which was later used in a modified and improved form by Henry Ford (see Chapter 3), has become the accepted design for tractors, and the Wallis curved steel plate model remained almost as a trademark for the firm long after they became part of the Massey-Harris organisation.

When the First World War brought new pressures and new opportunities to the tractor industry, manufacturing in America was stronger and more firmly based than in any other country, so it was well placed to supply the thousands of tractors required towards the end of the war to help put European farming back into full-scale production.

Britain

Tractor history in Britain dates back to about 1896, when a small number of pioneers began working on the idea of using an internal combustion engine to propel an agricultural machine. The first company to achieve commercial success with the idea was Richard Hornsby and Sons Ltd., of Grantham, Lincolnshire. Hornsbys had signed an agreement to manufacture and sell the oil engine developed in 1890 by Stuart and Binney, and this proved to be a successful venture. This engine, which was started by a blowlamp, preceded the engine developed in Germany by

Dr. Rudolph Diesel. Hornsby-Akroyd oil engines were particularly successful for agricultural work, and the demand encouraged the company to abandon their steam engine interests in order to concentrate on the internal combustion engine. Their engine was adapted to power a tractor, and in 1897 they exhibited the result at the Royal Show at Manchester, winning a Silver Medal award. The Royal Show model was fitted with an 18-b.h.p. engine mounted on a heavy steel frame consisting of two main members. The transmission was by gears giving three forward speeds, and there was a spur pinion to a pulley for stationary work. The price in 1897 was £500.

Charles Cawood, writing in the August, 1970, issue of *Industrial Archaeology*, records that one of these 'Hornsby-Akroyd Patent Safety Oil Traction Engines' was sold. The customer was H. F. Locke-King of Weybridge, Surrey. He bought it in 1897, so becoming the first person in Britain to buy a tractor.

The Hornsby was designed for haulage work and for operating stationary equipment, rather than for field operations such as tillage. In this respect, and also in its size and construction, it was more closely related to steam traction engines than to the concept of tractors which has become accepted. The judges conducting the trials for the Silver Medal award in 1897 obviously compared the Hornsby with steam power, and their report notes that the Hornsby saved the time required to supply coal and water to steam engines in the field. But the judges were also favourably impressed by the performance of the machine. They commented on its manoeuvrability and on its capacity to overcome obstacles, such as wooden sleepers, and to traverse soft, wet ground. Britain's first tractor was clearly a success on its first public appearance, and the manufacturers deserve more credit for their achievement than they have usually received. However, their tractor was doomed to commercial failure, not because it was too heavy or uneconomical—steam power remained a commercial success for decades after the Hornsby failed—but because it was too far ahead of its time. British farmers were simply not ready for the tractor, and it is significant that, ten years later, British manufacturers were relying on export sales rather than the domestic market to keep themselves in business. Richard Hornsby & Sons made a similar error a few years later, when they gambled heavily on the success of the crawler tractors they helped to pioneer, only to see failure in Britain while similar gambles were paying off in America.

While the Hornsby Company was introducing a tractor to compete primarily with steam traction engines, two men from Bedfordshire were starting work on much more advanced ideas for competing with the farm horse as well as the steam engine. These pioneers were Dan Albone and H. P. Saunderson.

Dan Albone's father was a market gardener near Biggleswade, Bedfordshire. Dan left the farm to start his own bicycle repair business, the Ivel Cycle Works, in Biggleswade. The Ivel was the river running close to Biggleswade and the name he later used for his tractors. The cycle business prospered, partly because of Albone's mechanical ability, and partly because of his talents as a businessman. Contemporary references suggest that he was a man of considerable energy, charm and cheerfulness, with an eye for detail.

His interest in tractor design probably dates from around 1896, but his development work did not reach what he regarded as a marketable stage until 1902, when he submitted his prototype tractor to the Royal Agricultural Society of England for assessment for their Silver Medal award. This first model had a two-cylinder horizontal engine developing 14 h.p. at 800 r.p.m. It had a single forward speed and reverse, each operated through an individual cone clutch. The total weight of the tractor, with full water tank, was 30 cwt., of which 22·5 cwt. was on the two rear driving wheels. Although the judges did not award the medal to the Ivel tractor until the following year, when some modifications were made, their report was full of praise for the tractor. The comments also show that the Society's judges were remarkably aware of the qualities which a farm tractor should offer.

The judges' comments, in the 1903 issue of the *Journal* of the R.A.S.E., include the following: 'Mr. Dan Albone ... has shown an intimate knowledge of what is wanted in an agricultural tractor, and has produced a machine adapted to rough handling and management by unskilled hands. The problem of making a tractor suitable for both field and road work is a difficult one, and can only be solved by a compromise, for weight on the driving wheels is essential for the hard road and lightness is an absolute necessity on the farm, particularly when the land is wet.'

The judges watched tests on land which was light, but soaked by heavy rain. Under these conditions the first Ivel tractor pulled three furrows of 8·75-inch width to a depth of 6 inches, at a forward speed of 3·5 m.p.h. The judges were concerned about the effect of tractor wheels in the furrow bottom, especially on wet, heavy soils—a matter still of concern seventy years later. They decided, however, that the light construction of the tractor, with only 11·25 cwt. on each driving wheel, would minimise the risk of damage to soil structure in the furrow bottom, and that the effect might not be as serious as that of a team of horses.

Albone succeeded in gaining financial support for the manufacture of his tractors from some of the most notable names in the contemporary business world. The 1904 model was equipped with two forward speeds and the power was increased to 20 h.p. This was the version which won the R.A.S.E. Silver Medal, and which Albone took to Paris for the 1904 farm machinery show, where it scored a remarkable triumph. According to one contemporary report it was the centre of attraction. The fact that Dan Albone exhibited at an overseas show so early in his business career indicates how seriously he rated the export market for his tractor. Overseas sales were already an important part of his business, and by the end of 1904 there were already Ivel tractors working in many countries, including South Africa, Egypt, Portugal and France. In 1904 he built a modified, low profile version of his tractor to meet a special order from a Tasmanian fruit grower who wanted to work between the trees in his orchards.

Dan Albone's salesmanship and eye for publicity were used to good effect. A journalist visiting his Biggleswade factory was met at the station by a car. This was a sufficiently notable happening in 1903, when cars were a rare luxury, for the journalist to comment on it in his subsequent report. By 1903, Albone already had operators' instruction manuals available for his tractors, and had purchased land near the factory for tractor tests and development.

Ivel tractors were demonstrated at trials and tested before independent witnesses to record working rates, fuel consumption and operating costs. In one of these tests in Bedfordshire in 1903, an Ivel tractor cut six acres of meadow grass in three hours and forty minutes, consuming 5·5 gallons of petrol and one pint of lubricating oil. The petrol cost 7s 4d (36½p) at 1s 4d (7p) per gallon, and the driver's time, at 1903 rates of pay, was calculated at 1s 9d (9p) for the three hours and forty minutes. The man on the mower earned only 12d (5p) for the same time.

The contribution which Dan Albone made to the development of the farm tractor was enormous. He died in 1906, but in only about ten years of development work, including five years of commercial experience, he introduced and energetically promoted a concept which was completely revolutionary. He was the first person to develop and market a tractor which was both small and efficient—the forerunner of the tractors which millions of farmers would eventually buy. His name is scarcely known outside the ranks of tractor enthusiasts, but if he had lived long enough to develop his ideas further and to take advantage of improvements in engine design, the name Albone would be as familiar to farmers throughout the world as those of Ferguson and Ford.

After Albone's death, the Ivel company continued to manufacture and market tractors, and a slightly modified version of the Ivel was produced in America by the Russell company. But the company appears to have lost some of its business flair, and development work was not active enough to keep the Ivel in the forefront of design. The firm did not survive the war, and went out of business with less than one thousand tractors built.

H. P. Saunderson was another outstanding inventor, and Saunderson tractors became the best-selling British make before the First World War. As with the Ivel, development work started in the closing years of the nineteenth century, but nothing appeared commercially until 1904 when the extraordinary horse-substitute, described in Chapter 2, was shown. It is significant that this machine was intended as a substitute for a horse, rather than for a traction engine with which almost all of Saunderson's contemporaries were competing. The following year a more powerful version appeared, with provision for fitting a truck body to the chassis— further evidence of Saunderson's far-sightedness. This tractor was apparently fitted with Ackermann steering, and in the 1905 Royal Show Saunderson continued his pioneering development by exhibiting and demonstrating a version of the tractor with four-wheel drive. By 1906, when the Saunderson eventually won a Silver Medal at the Royal Show, the tractor had been up-rated to 30 h.p. and was spring-mounted on its three wheels.

Saunderson tractors eventually achieved commercial success, both in

Britain and overseas, with the basically conventional Universal models, which sold in substantial numbers. Many of these are still in existence in museums and collections.

Before the First World War Britain's tractor industry, which consisted mainly of Saunderson, Ivel and Marshall of Gainsborough, found the home market generally unreceptive, and it was export sales, especially to countries in the British Empire, which provided business for the factories. Other companies, and individuals, such as Drake and Fletcher of Kent, Petter of Yeovil, Ransomes and Sharp, which introduced prototypes in this period, were presumably discouraged by the British farmer's apparent determination to work with horses or steam.

Europe

There was, on the whole, little commercial development of tractors in Europe until war-time conditions created a substantial market for farm power. However, in both Germany and France there were outstanding exceptions to this generalisation.

An early German development, which is interesting even if of only local significance, was the electric ploughing system developed by A. Borsig of Berlin. According to an account published in the British journal, *Implement and Machinery Review* on 3rd July, 1897, Borsig's ploughing system was in commercial use in 1894. It was developed specifically for the sugar beet producing areas of Germany, where farms were extensive and highly profitable. The sugar beet processing factories in the beet growing areas used steam to generate electricity, and at certain periods had surplus power available. Borsig's system used this spare current to power motors which were mounted on four-wheel chassis. The motor units were self-propelled, and could be driven from field to field, towing the plough and an anchor unit in a little train behind them. In the field, the motor unit was positioned on the headland and operated the plough by means of a winch and cable system. These motors were powerful enough to operate a three-furrow balance plough. The value of the sugar beet crop, and presumably the low cost of the locally produced off-peak electricity, justified the cost of taking power cables to out-lying fields on

16

the estates. The report in *Implement and Machinery Review* implies that the system was widely used.

More orthodox German development of tractors came from Deutz, a company as old as the four-stroke engine. The first Deutz tractor was built in 1907 and designed specifically as a ploughing tractor. A second, improved version that appeared in the same year, seems to have been the first genuine two-way tractor. Contemporary photographs show this tractor with a central, upright steering wheel, and two seats arranged so the driver could sit on either side of the wheel. The tractor was supplied complete with two four-furrow ploughs, which were attached fore and aft. Both ploughs were mounted on wire cables, with provision for raising them out of work. To operate this system, called the System Brey, the driver sat behind the wheel, raised the plough in front and lowered the one behind, and drove forwards. At the end of the field, he moved to the other side of the wheel to face the way he had come, raised the plough now in front of him, lowered the one behind the tractor and drove back towards his starting point. The tractor, on four equal-sized wheels, was fitted with a 40-h.p. engine.

The Deutz two-way tractor plough unit was apparently not a great commercial success, and the same fate befell the ideas of two French pioneers, Gougis and Bajac, who were also working on tractor development in 1907. Gougis made a prototype tractor equipped with a four-cylinder engine, and with a power-take-off shaft at the rear for driving trailed machinery. He developed the idea for a binder, and it appears to have been successful. However, Gougis, for some reason, failed to develop the idea of power-take-off commercially, and it was International Harvester who reputedly saw the Gougis machine at a French show, recognised its real potential and introduced the p-t-o on one of their own tractors.

Bajac developed the idea of a compact tractor for inter-row cultivations, which he fitted with a range of mounted implements with an arrangement for manually raising them out of work. Neither of these two French pioneers is known to have made any money from inventions which eventually became almost standard features of tractors throughout the world.

TESTS AND TRIALS

Technical improvements in tractor design and the commercial acceptance gained during the period before the First World War increased the hostility of manufacturers of steam engines and breeders of farm horses. This hostility seems to have been particularly evident in North America, and horse breeders were still fighting a rearguard action against the tractor well into the Second World War.

The commercial conflict included a war of words which must have been most confusing for the farmer trying to assess the merits of the three power sources. The confusion was aggravated by the numerous biased costings which were produced by supporters of these rival power sources. At the same time, tractor manufacturers competing against each other in a market where customers were rarely well informed, were sometimes tempted to make extravagant claims about performance.

In this situation a few far-sighted people realised the need for farmers to have a basis for objective assessment. If various makes of tractor could be seen working side by side in the same field conditions, with careful measurement of work done and fuel consumed, farmers would be able to choose more effectively. The first major opportunity for tractor manufacturers to compete on level terms in public was the Light Agricultural Motor Competition held in 1908 in Winnipeg, Canada.

The Winnipeg competitions

The man who did most to get the Winnipeg competitions off the ground was A. Burness Greig, lecturer at an agricultural college and a keen advocate of the farm tractor. He was appointed engineer in charge of the 1908 competition, and was closely concerned in the organisation of the event in subsequent years. A feature of the competitions, which were held annually until 1914, was the thorough and detailed way they were planned to assess and compare the machinery taking part. Much of the credit for this high standard should go to Burness Greig.

The competitions were organised in conjunction with the Winnipeg Industrial Exhibition. The regulations of the first competition limited the weight of competing tractors to seven tons, a level which excluded steam engines. Eight tractors were entered in 1908, of which one was above the weight limit and another was withdrawn after a mechanical failure. The six which completed the course were a Kinnard-Haines, a Transit Big Four and three Internationals, all from America, and a Marshall imported from England, which was the only kerosene or paraffin burning tractor.

After the tractors had been weighed, they were put through a drawbar test, in which they pulled wagons loaded with gravel over a measured course. The Transit and the Kinnard-Haines both pulled two trailers with a total weight of 18,040 lb. and the others took one trailer each with approximately half the weight. These loads were pulled for approximately two hours, during which time most of the tractors managed to travel up to six miles.

For this section of the competition efficiency was measured by comparing the work done in foot-pounds with the quantity of fuel used. The Kinnard-Haines used nearly ten gallons of petrol to pull its two trailers slightly more than five miles. The judges also measured the quantity of water used by the tractors during this test, and the results show the limitations of the early hopper and open-top tank cooling systems. The Marshall tractor started the course with seventy gallons of water and at the finish had lost eighteen gallons through evaporation or leakage. The Kinnard-Haines used only six gallons of the 164 gallons it started with, and the Transit apparently used no water at all during the test. The three International tractors used the most. The three-cylinder 40-h.p. model, the most powerful of the tractors competing, used thirty-two of the forty-four gallons in its tank.

Two days later, the tractors' performance with ploughs on heavy, rain-soaked land was assessed. Typical of the tests' attention to detail and fairness was the fact that all the competitors were provided with Cockshutt ploughs to eliminate differences of furrow design which might affect efficiency. Each entrant chose the number of furrows he wished to pull, and the judges measured the area ploughed and the quantity of fuel used in approximately two hours. The Marshall tractor, which had done well in the drawbar tests, performed badly in the ploughing section. It took eighty-one minutes to plough an acre with three furrows and used fuel at the rate of thirty-eight pints an acre. The Kinnard-Haines, which was rated at 30 h.p., the same as the Marshall, pulled six furrows and took thirty-seven minutes and twenty pints of fuel per acre ploughed.

A complicated system of points was used to calculate the overall result, which was designed to measure working efficiency in the two sections of the competition, rather than simply the total amount of work done. The Kinnard-Haines four-cylinder tractor, weighing 13,530 lb. was awarded 117·6 points, to win the competition by a margin of 0·6 points over the International single-cylinder 15-h.p. model, which at 9,920 lb. was the lightest tractor to complete the competition. The Marshall earned 108·3 points to take third place, followed closely by the 40-h.p. International, the 35-h.p. Transit and the 20-h.p. International.

The 1908 event was rated a success, and the organisers planned the next year's competition on an even more detailed and ambitious scale. The seven-ton weight limit on entries was removed, and five steam engines competed in a class of their own. Tractor entries were divided into three classes. There were three entries in the class for tractors of less than 20 h.p., and five each in the classes for 20 to 30 h.p., and 30 h.p. and above. International had seven tractors competing, again the largest entry. They won the classes for the small and medium power tractors and took third place in the remaining tractor class. Kinnard-Haines won the section for tractors of 30 h.p. plus and a Marshall was placed second.

The results' sheet for 1909 was extremely informative. It quoted a mass of statistics, including details of the total weight and weight distribution of each entry, cooling and fuel tank capacities, diameter of driving wheels, pulleys and the price. The 25–60-h.p. Marshall at 22,000 lb. was the heaviest tractor competing, and at $3,400 was also the most expensive. If there had been a prize for the biggest water tank, it would have been won by the Kinnard-Haines 40–60 with a capacity of 290 gallons.

In later competitions points were awarded for the design and construction of competing tractors, as well as for their performance and efficiency. In 1912 accessibility, protection of working parts and ease of manipulation were all taken into account. The records for that year also show the penalties imposed for various infringements of the rules: one entrant lost points for talking to a colleague during a test, and another incurred a penalty for setting his plough furrows deeper during the dynamometer tests.

The Winnipeg competitions were discontinued in 1914, apparently from lack of financial support. But by then there were many different tests for manufacturers' implements, which enabled farmers to compare performances and to see how manufacturers' advertised claims matched up to reality. Burness Greig and his colleagues set a really worthwhile standard of objective assessment in their competitions, and farming has much to thank them for. The Winnipeg competitions, and similar events, helped bring tractors to farmers' attention at a time when steam power and horses set the accepted standards. The competitions, by publicising the weaknesses of inferior designs, also encouraged the manufacture of better tractors.

R.A.S.E.

The Royal Agricultural Society of England is one of the oldest and most respected farming organisations in the world. Its function is to encourage farming progress, especially in the adoption of improved scientific ideas and techniques. In its long history the Society has done much to evaluate and promote developments in mechanisation, and this applies particularly to tractors. Better tractor design and utilisation have been fostered in several ways. The Society published authoritative papers dealing with aspects of tractor use, organised trials to evaluate and compare different makes and models, and awarded Silver Medals to manufacturers introducing what the Society's judges considered to be worthwhile advances in design or specification.

R.A.S.E. Silver Medals, awarded annually at the Royal Show, are a distinction which must be earned. Machines entered for awards are tested by the judges, and over the years the tests have gained an enviable reputation for their thoroughness. The first tractor to earn a Silver Medal was the 'Hornsby-Akroyd Patent Safety Oil Engine, Traction or Agricultural Locomotive'. The award was made in 1897 when the Royal Show was held at Manchester.

The value of the Silver Medal awards was demonstrated a few years later when two tractors were turned down by the judges, but invited to enter again if certain improvements were made. The first of these was Dan Albone's Ivel tractor. It gained a Silver Medal in 1904 after the single speed with which it had been entered in 1903 had been replaced by two forward speeds, and the cooling system improved.

A Saunderson tractor, entered for an award in 1905, failed to meet the judges' standards. It reappeared the following year, at the Royal Show, Derby, and the improvements incorporated in the design earned it a Silver Medal. This tractor was an unconventional but ingenious design. It had a 30-b.h.p. engine which was spring mounted, and the tractor could be used in three different ways. In its standard form it was a three-wheel drive haulage unit or stationary power unit. A truck body could be attached for carrying loads up to a claimed maximum of four tons. The third adaptation was as a forecarriage on to which specially modified implements, including ploughs or drills, could be directly attached. This tractor appears to be the original version of much more recent ideas of the Unimog or Fendt toolcarrier type.

The fourth tractor to gain a Silver Medal, awarded in 1915, was the Walsh and Clark Victoria ploughing engine from Yorkshire. The design of this machine was a close imitation of steam traction or ploughing engines. The enormous tanks, shaped and located like a steam engine boiler, carried enough fuel oil and water for twelve working days.

In 1924 a Silver Medal went to the extraordinary Fowler Rein Controlled Tractor. Although the design was not a commercial success, the judges were enthusiastic about the idea. The driver sat or stood on the implement which was being pulled, and worked the tractor controls at long range by means of two reins. 'A slight jerk of the reins puts the tractor into forward gear, and the band clutch takes up the drive quite smoothly,' said the judges' report. 'A slight pull at the left or right rein steers the tractor. A steady pull on both reins stops it and a further pull puts the reverse gear into action. The reins are easily operated by one hand, leaving the other free to attend to the implement being drawn. The power steering enables the tractor to turn over into full lock while standing.'

If this tractor, which was apparently equipped with power steering years before it was available elsewhere, and had a twin-cylinder, V-design engine developing 30 b.h.p., really was so easily controlled, it is surprising that it failed to sell. It was ideal for the thousands of farmers with serviceable horse-drawn equipment available, and drivers used to working with horses would have felt at home operating the reins.

The next Silver Medal was awarded in 1928 to the French Latil tractor, and the judges commented favourably on its advanced design and novel features. In 1931, when the Royal Show was at Warwick, an award went to a diesel tractor built by Garrett of Leiston, Suffolk. The engine on this tractor was a four-cylinder, four-stroke Aveling and Porter unit, rated at 23–40 h.p., and equipped with electric and hand starting. The tractor had three forward speeds and one reverse.

The R.A.S.E. judges particularly welcomed the engine of this tractor, realising its advantages at a time when high-speed multi-cylinder diesel engines in tractors were extremely rare in Europe, and unheard of in America.

According to the report in the R.A.S.E. *Journal*, 'The diesel engine is, of course, far superior to the petrol or paraffin engine as it is simpler, there being no magneto, sparking plugs, carburettor or vaporiser, and also there is no excessive dropping of power at reduced engine speeds. These four-cylinder engines, with their high speeds and even torque give that steady drive which is essential to good ploughing. To sum up, this tractor may be described as a remarkably fine piece of English engineering workmanship that should outlast the average tractor. It was a pleasure to an engineer to look at.'

More recently, Silver Medal awards have gone to the Doe Triple-D tractor, an articulated design using two Fordson Super Major engines to give four-wheel drive, and to Ford, and to Massey-Ferguson for pressure control and for multipower transmission. In 1972 Massey-Ferguson received an award for their 1200 tractor. Although this was primarily for the articulated four-wheel drive design, the standard of comfort in the cab was also much praised.

The Royal Agricultural Society's tractor trials started in 1910. In 1898, a trial for self-propelled vehicles attracted a Saunderson petrol engined machine, but it appears that this was more of a truck than a tractor and it failed to complete the course against steam competition.

The 1910 trials were held near Baldock, Hertfordshire. They were intended to compare the performance of tractors and steam engines. Preparation for the event was thorough, and so was the report of the trials in the R.A.S.E. *Journal*, which ran to twenty pages, including six pages of tables. The tractor entry was disappointing, and only two Ivels and two Saunderson Universals took part, against three steam traction engines. All entries had to plough light land and heavy land, have their power measured on a dynamometer, haul loaded trailers twice round a twelve mile course covering both smooth and rough surfaces, and pull a pair of reaping machines in wheat. Fuel consumption was noted throughout, as were breakages, and the judges had yard square samples of soil removed to ploughing depth from all the plots and weighed in order to calculate the weight of soil moved per acre. Ploughs, trailers and reapers used by the entrants had to be of the same make and design.

Fuel consumption for ploughing was recorded for all the entrants. A McLaren steamer used only 54·7 lb. of coal per acre, but the Mann traction engine used 96·5 lb. Paraffin consumption was 3·63 gallons per acre for the single-speed Ivel, 3·36 for a two-speed Ivel, 4·44 gallons for the 25–30-h.p. Saunderson and 5·16 gallons for the 45–50-h.p. model of the same make. None of the tractors performed well enough to satisfy the judges, and no awards were made. However, the published report on the trials expressed the conviction that the tractor offered the best hope for a completely general-purpose farm power unit, and hoped that the trials would stimulate the development of such a machine.

Plans for further trials in 1915 were delayed by the war, and the next R.A.S.E. event of this type was the 1920 Motor Trial at Lincoln. This, too, was an ambitiously organised affair. The Society set aside eight hundred acres to test the tractors, and allocated sixty-two pages of the *Journal* for the report, with eleven pages of photographs. The prize list shows clearly the extent of the American influence which had appeared during and just after the First World War. Five tractors won awards. These comprised the Peterbro, which was completely British, the British Wallis of American design but British manufacture and the Case 10–18, Cletrac and Lauson which were all imported from the States. There were also classes for steam traction and ploughing engines, for cable ploughing engines with internal combustion engines and for motor ploughs. In the motor plough class a British Crawley model won, with a Moline in second place.

Even more ambitious were the World Tractor Trials which the R.A.S.E. organised in 1930 with the help of Oxford University. In the tractor classes there were thirty-three entries from seven countries. The testing of the entries began on 2nd June and ended on 26th July, with a public demonstration in September. These trials were intended to give a thorough testing and to provide a mass of data on the tractors, they were not competitive. The event drew tractors from virtually every manufacturing country in the world, except Russia. It provided a rare opportunity to compare products from Britain and America with those of less familiar manufacturers, such as H.S.C.S. in Hungary and companies in Sweden. The Fordson represented the Irish Republic, where it was then manufactured.

The Nebraska tests

Rapidly increasing demand for tractors during the First World War led to many new makes and models appearing on the United States market. Often these tractors were badly designed and manufactured, and were backed up by completely inadequate service facilities. As the situation deteriorated, the demand grew for some programme of assessment for tractors, based on independent testing with standardised procedures.

This demand came from the more responsible sectors of the tractor and farm machinery industry, as well as from aggrieved farmers. At first little progress was made, largely through lack of finance, but the breakthrough came in 1919. In that year the Agricultural Engineering Department of Ohio State University conducted a series of ploughing tests on a wide range of soil types, and legislation was introduced and approved which lead to the famous Nebraska tests.

The Nebraska Tractor Bill was brought to the State Legislature by Representative Wilmot F. Crozier. He was a farmer's son from Polk County, Nebraska, and divided his time between working at home on the family farm, politics and teaching in schools in various parts of the United States. His ideas about the need for proper testing were probably the result of experience on the family farm with various makes of tractor which showed marked differences in performance and reliability.

Crozier's Bill was aimed at preventing a manufacturer from selling a tractor in Nebraska on the basis of false claims or with inadequate service facilities. Under the Nebraska Tractor Law it became an offence to sell a tractor in the State unless a permit had been obtained. This permit could be issued only after an example of the same make and model had been officially tested, and its performance during the test had borne out the manufacturer's claims. The permit also depended on the provision of approved service facilities.

The Agricultural Engineering Department of the State University was to be responsible for conducting the tests. A standardised testing procedure was devised and suitable equipment was developed ready for testing to begin early in the winter of 1919. The first tractor to be submitted to the test programme was a 12–20 Twin City, but unfortunately weather conditions were too bad and snow made it impossible to continue with the drawbar tests until the following spring. The first tractor actually to complete the full programme was a Waterloo Boy Model N, and a further sixty-four tests were completed during 1920.

The results of these early tests provide an excellent record of performance for the tractor historian. The Fordson tested in 1920 was distinguished by its comparatively light weight—2,710 lb.—and by the 23·80 per cent wheel slip recorded in the drawbar tests. The only other tractor to exceed twenty per cent wheel slip was the tiny Beeman Model G which developed a little over 2 h.p. in the belt tests and weighed only 550 lb. Heavier tractors achieved significantly better results when wheel slip was measured, and the 60 h.p. Aultman-Taylor, weighing 24,450 lb. which is slightly heavier than a modern 320-h.p. Steiger tractor, recorded only two per cent slip.

The 1920 tests showed marked variation in fuel consumption between various tractor models. The Fordson returned a decidedly moderate figure of 6·38 horse-power hours per gallon of kerosene or paraffin, during the maximum load belt test. In the same test the big Aultman-Taylor gave a figure of 7·46, and a 15–27 Case was outstandingly economical with 9·90-

h.p. hours per gallon. The Uncle Sam 20–30, with a fuel figure of 5·02, had the largest fuel consumption. However, it used only ·14 gallons of water per hour in the maximum load test. Water consumption was generally two or three gallons an hour, but the 24–50 Avery boiled, or possibly leaked, twenty-five gallons an hour, with a 40–80 tractor of the same make using almost nineteen gallons an hour.

Results of the Nebraska tests were widely publicised and became accepted as the yardstick for comparing tractors throughout the United States, and later, in many other countries. Farmers could take the results into consideration when buying equipment for their own farms, a fact shown by the manufacturers' sales figures. Poor test performance led to falling sales, which had to be remedied. Thus the tests have done much to improve design and performance standards and to discourage the inferior. Crozier's efforts to protect the interests of Nebraska farmers have become the most internationally respected assessment of tractor performance.

3

THE FIRST WORLD WAR

It is difficult to avoid the conclusion that the First World War generally helped the tractor industry and provided a considerable stimulus to technical development. Before 1914 there was a well established tractor industry in North America, and to a lesser extent in Britain, but few firms were involved on a commercial scale in Europe.

The appalling loss of life on the battlefields of Europe and the devastation of years of fighting created a serious shortage of food. The war effort required huge numbers of horses and mules to move material to the front lines. Tremendous numbers of these animals were killed, seriously reducing the food production capacity of agriculture. A substantial proportion of Europe's farming land became battlefields during the war years, with serious damage to buildings and fences, and loss of crops and livestock. The German U-boat campaign during the war effectively isolated Britain from much of the vitally needed food imports from across the Atlantic.

In this situation it was fortunate that American manufacturers had the capacity and the raw materials to increase tractor production. Tractor power was needed in America to increase food supplies for export to Europe, and, towards the end of the war, to make good losses of manpower and farm animals for military purposes. Tractors were desperately needed in Europe to provide the power to put farming back on to a productive basis.

During 1917 tractor production in the United States approximately doubled to reach almost 63,000 units. Exports, mainly to Britain and other European countries, accounted for 15,000 of these, and more than eighty new companies entered the market in that year alone. Production in 1918 doubled again, to more than 130,000. The high level of demand at this time provided an opportunity for considerable experimentation in tractor design. From the abundance of new ideas there emerged a limited number of worthwhile developments, and also an extraordinary miscellany of complete failures.

Tractor designers were particularly interested in methods of improving drawbar pull. Tractors were marketed with conventional two-wheel drive, and also with four-wheel, three-wheel or only one-wheel drive. The 1919 Victor tractor consisted mainly of two very large front driving wheels, with the engine mounted inside the wheels and transmitting power through two sets of gear teeth arranged around the inner circumference of the wheels. Steering was by a single rear wheel, only a little more than one foot in diameter. The Post tractor, produced in America at about the same time, had its four wheels arranged in a diamond formation. The front and rear single wheels both drove and steered and two idler wheels, arranged as outriggers, prevented the Post tractor from falling over sideways.

However, American designers were also producing sound designs with considerable technical improvement, which moved away from the concept of massive size and weight inherited from the days of steam power. Many of these tractors were exported, particularly some of the Mogul and Titan models, the International 8–16, the Moline motor plough and, of course, the Fordson. These tractors were of a size suited to the scale of European farming, and were relatively reliable and long-lasting.

These and other American tractors gave many farmers in Europe their first experience of tractor power, a factor which gave the American industry a strong position in the developing European markets. Well-tried American designs, with prices based on production volumes which no European manufacturers could achieve in the immediate post-war years, proved effective competitors for the new models being introduced in Europe.

The situation can be judged from a report published in the R.A.S.E. *Journal* describing the operation of the tractors owned for war-time food production by the Kent War Agricultural Committee. Fifteen makes of tractor had been tested in that county alone, and at the end of the war the Committee in Kent owned and operated 180 tractors—112 Titans, 4 Overtimes and 64 American Fordsons. These tractors all earned favourable comments from the Committee, which had kept careful records of hours worked and fuel consumption, although the Fordsons appear to have shown significant advantages in work rate and fuel economy compared with the other makes.

For many years before the war, Britain had been the leading exporter of tractors, and had earned an excellent reputation for good design in many countries. After the war the British industry was in a weak and poorly organised state, with a home market dominated by the American makes which had gained such an effective war-time foothold and reputation.

Tractors and tanks

Farmers and soldiers both require vehicles which can pull or carry loads over rough ground. During the First World War several attempts were made to adapt tractors for military use, and after the war there were some more successful attempts to adapt military vehicles for agricultural use.

The Ivel tractor was used as the basis for an experimental armoured vehicle as early as 1910. In his book, *Tanks and other Armoured Fighting Vehicles, 1900–1918*, B. T. White records that army interest in the Ivel was probably due to its potential as an artillery tractor. The Royal Marines were apparently interested in using the Ivel for ambulance work, presumably for towing a trailed ambulance. These ideas seem to have been dropped after the prototype Ivel in armour had been produced.

Tracklaying agricultural tractors had considerable impact on the design of war vehicles. The British army adapted the Holt Caterpillar tractor from California for use as an artillery tractor for moving heavy guns, and they also imported other makes of American crawler tractors for the 'Landship' project. This was a proposal for a vehicle to travel across the trench system of the First World War battlefields, which was creating such deadlock in the fighting. The American tractors involved were the Bullock Creeping Grip from Chicago and the Killen-Strait produced in Appleton, Wisconsin. The Bullock design was fairly advanced and has since become conventional, with two tracks almost the full length of the tractor.

It was the Killen-Strait which had the greater long-term impact, in spite of its rather peculiar layout. The three tracks arranged in tricycle form, the front track for steering and the rear pair for driving, would have been quite unsuitable for the army's main objectives. However, the tractor performed well in tests before Winston Churchill and other war-time leaders, and it was later fitted with an armoured body shell to become the first tracklaying armoured vehicle built and a direct ancestor of the army tank. William Strait, who held the patents to this tractor, had established the idea of having the driving tracks tilted up at the rear to enable the vehicle to be reversed over obstructions. This was demonstrated to British Government and army leaders in the summer of 1915 and later appeared on most of the world's tanks.

After the war had ended there were some successful attempts to convert military vehicles to farm use. The Daimler Company, a forerunner of the Austrian Steyr-Daimler-Puch organisation, produced what they called a 'horse substitute' in 1917, which appears to have been an artillery tractor with two wheels and 4-cylinder, 14·5-h.p. engine. This was certainly adapted for tillage work after the war and was a predecessor of the present Steyr range of tractors. Renault produced a particularly successful light tank during the war, which was manufactured in large numbers. In 1918 the Renault GP tractor was announced, based directly on the Renault tank, but with a distinctive body style which remained in production until about 1930. The H1, which launched Renault into a long and successful involvement in tractor manufacture, was equipped with a 30-h.p. petrol engine, and had three forward and one reverse speeds operated through a cone clutch. The H1 was a tracklaying model, but wheeled versions of the same basic design appeared later.

One German machine which appears to have earned a particularly unfortunate reputation in Britain was the Stock Motor Plough from Berlin. The manufacturer had exported one of these to Britain before the outbreak of war, and this turned up, rather surprisingly, at the Royal Show held at Shrewsbury soon after the war started in 1914. The R.A.S.E. *Journal*, in its description of machinery exhibits at the show, made the following comment on the Stock Motor Plough:

'Visitors to the Show, who saw this exhibit and noticed its enormous wheels and great power, will be interested to know that, in all probability this is the identical type, if not the actual machine that has been used by the Germans for digging their trenches and burying their dead.'

One of the benefits which the First World War brought to the tractor industry was the might of Henry Ford's manufacturing resources. He chose an excellent design and put it into production on a scale and at a price which gave a considerable impetus to the pace of farm mechanisation, as well as challenging competitors.

The impact of Henry Ford, like that of Harry Ferguson, survives in most of the tractors in use today, and these two men have other things in common. Both were outstanding engineering pioneers—but it was said that neither could understand a blueprint. Ferguson's father farmed in northern Ireland. Ford's father left southern Ireland to farm in Michigan. Both Ford and Ferguson left the farm as soon as they were old enough to earn a living in the developing engineering industry, and both became wealthy men.

Henry Ford's father sailed from Cork after the Irish potato famine of 1846 to join relations already established in Michigan. He cleared forest land to create a farm near Detroit, where Henry Ford was born in 1863. Childhood on the family farm appears to have given Ford his first ideas for using mechanical power to replace the slow pace and drudgery of working with horses. He disliked the limited scope of farm life, and left home at the age of seventeen to gain industrial experience in a series of jobs in Detroit.

Designing engines was one of Ford's early interests, and he was experimenting with ideas on the Otto principle in the late 1880s. By about 1892 he had built his first successful engine, and in 1896 he completed and drove his first car or quadricycle. This had two forward speeds and no reverse, and was driven by a belt which was later replaced by gears. Henry Ford succeeded in gaining the financial backing he needed to develop his cars commercially, and he and his partners formed the Detroit Automobile Company in 1899. This first company had a short existence, and it was succeeded by the Ford Motor Company in 1903.

In his early years as a manufacturer, Henry Ford concentrated his efforts on cars, but he claimed to have been interested in tractor design even before he first became involved with cars. He designed and built his first prototype tractor in 1907, and this model is preserved in the Henry

Ford Museum at Dearborn. It was built mainly with car components, and included a Model K radiator, front axle and steering, and the engine, which was mounted transversely, was the Model B 20-h.p. unit with four individual cylinders. The idea of using components already in volume production for Ford cars, appealed to a man so aware of the cost advantages of mass production. Several experimental and prototype tractors were produced which included parts of other vehicles. These included a 1936 tractor with Ford truck engine, radiator and modified steering, and car wheels in the front, and there were also several tractors based on the Model T car.

The Model T Ford is probably the most significant car in the history of the motor industry. It was launched in 1908 and continued in production until 1927. During this period the Ford Motor Company produced 15·5 million Model Ts, bringing the convenience of car ownership for the first time to millions of families. The economic and social importance of the Model T is enormous, and many books have been written about the car which dominated the motor industry for so long, and which helped to introduce a completely new scale of manufacturing.

The extraordinary success of the Model T was the result of Henry Ford's ideal of producing a sturdy, simple, reliable car at a price within the reach of a vast new market of would-be car owners. The car itself was remarkable, but even more remarkable was the vision and achievement of Henry Ford in foreseeing the massive potential of a car produced in huge numbers at low cost. His entry into the tractor market shows a similar approach, and the Model T was in many ways a pattern for the Fordson tractor. The tractor, like the car, was sturdy, simple and reliable, it was planned for large-scale production and priced at a level which put tractor ownership within reach of many farmers for the first time.

Henry Ford announced his plans for manufacturing tractors on a commercial scale in 1915, and he formed a new corporation, Henry Ford and Son, to design and produce them. Thanks to the success of his Model T car, Ford could back his new project with enormous resources of money and skill. Some of his best designers were set to work on the tractor, and one of these, an Hungarian named Eugene Farkas, developed the idea of using the engine, gearbox and rear axle castings to form a rigid structure strong enough to make a separate steel framework unnecessary.

The 1915 Fordson prototype was not the first tractor designed without a conventional frame. The Wallis Tractor Company of Cleveland, Ohio, had produced their Cub model two years previously, in which the backbone of the whole tractor was a steel U-frame which served both as a frame and an enclosure to protect the working parts of the engine and transmission. The Wallis Company became part of the Massey-Harris organisation in 1928, when Massey-Harris acquired the J. I. Case Plow Company of Racine, Wisconsin. The steel U-frame was retained in Massey-Harris models such as the Pacemaker and Challenger until the late 1930s.

Tractor conversions for the Ford Model T car were popular in America and Britain before the Fordson superseded them.

The U-frame was an excellent design, and provided considerable strength. But it was expensive to manufacture, and it was the Farkas design of stressed cast iron units which most other tractor manufacturers have since tended to follow. After some persuasion, Henry Ford agreed to try the idea of frameless construction, and a pre-production run of about fifty tractors was authorised for field testing.

Although Henry Ford had taken a great deal of interest in the tractor project, he appears to have been in no great hurry to begin volume production. Development work continued, however, and Ford waited for the improvements he considered necessary. Ironically it was the war that finally pushed Ford into giving the go-ahead for production to begin. Henry Ford had taken a firm stand against war and especially against American involvement in the First World War. He had proclaimed his views publicly in terms which he was probably to regret, and he also financed the famous peace ship crusade, which he may also have regretted.

The peace ship was a liner which Ford chartered to carry himself and a group of leading pacifists to Europe with the praiseworthy but rather impractical intention of ending the war.

Ford's decision to put his tractor into production in 1917, even though he considered it still imperfect, was prompted by urgent appeals from Britain. Several years of war had proved a serious strain on British resources, particularly food supplies. German U-boats were seriously limiting the supply of imported food, and the loss of manpower and horses from farms for service on the battlefield had made it difficult to maintain food production. The situation was becoming critical, and tractor power was urgently needed to help increase Britain's arable acreage. Companies in Britain which might have been able to produce the tractors were already involved in armaments, and there was no British tractor available which had been designed for large-scale production. Two of the Fordson prototypes had been imported by the British Government for testing, and their performance and their suitability for British conditions made a favourable impact. Henry Ford received a request to authorise production on a Government-controlled basis in Britain, and later, when this proved impractical, there followed a plea that he should manufacture the tractor immediately in the States against a substantial first order from Britain.

Henry Ford, who had dropped his stance against involvement in the war, had already agreed to put his factories behind the war effort and was either producing or planning to produce characteristically large numbers of aircraft engines, ships and tanks. He quickly agreed to the British request for tractors, and the Fordson went into production in 1917, with the first seven thousand shipped to Britain within six months. Some of these early Fordsons were still working in Britain after the end of the Second World War, and a few have been preserved in museums and collections.

The name Fordson may have been used simply to identify the tractor as a product of Henry Ford & Son. This was a separate company wholly owned by members of the Ford family, while the Ford Motor Company was partly owned by a small number of shareholders. The American authority, R. B. Gray, in his book *Development of the Agricultural Tractor in the United States*, gives a different reason for avoiding the name Ford, which is one of the most widely recognised trade names in the world. He quotes H. E. Everett of *Implement and Tractor* who claims that when the news broke that Henry Ford was developing a tractor, a group of Minneapolis business men hired a man by the name of Ford so that they could register the name for a tractor they produced. They launched the Ford tractor, hoping that some farmers would associate the name with Henry Ford and the Model T car. Certainly there was a Ford tractor manufactured by a company which had no association with Henry Ford, and this could well have influenced the decision to call the Dearborn tractor Fordson.

The Fordson was notable for several features besides the cast iron units to replace a frame. It was a compact, lightweight tractor, and it was marketed at a price which helped to put many other manufacturers out of business. Henry Ford stuck to his policy of selling the tractor cheaply, even when he could have exploited the strength of the tractor's position in the market by marking up greater profit margins. In his determination to sell a good tractor as cheaply as possible, Henry Ford's policy was similar to Harry Ferguson's. Ferguson believed that his tractors would revolutionise world farming, but recognised that they had to be cheap enough to be purchased for small acreages.

Henry Ford is one of the outstanding people in industrial history. Anybody who rises from a modest farm background to achieve complete control of a $500 million corporation is unlikely to be ordinary. Although many books have been written about him, few give much more than a passing reference to his contribution to tractor development, mainly because it was a relatively tiny part of his total achievement. His power and wealth, and the way he acquired and used them, his Model T and other cars, plus the extraordinary success of the Trimotor commercial plane, are all of greater interest to most Ford historians. However, tractors were involved in one of the most extraordinary of Henry Ford's activities—his support for the then recently victorious Communist Government in Russia. Henry Ford, one of the most powerful men in the capitalist world, gave much support to Communist Russia, and his motives were certainly not primarily the profit he could gain.

Russia in the early 1920s desperately needed machines and technology, and Ford was well placed to supply both. By 1926 Russia had purchased

25,000 tractors from Ford, who, in 1927, could claim that eighty-five per cent of the trucks and tractors in Russia were his products. In 1926 the Russian Government invited Ford to send a delegation to see how their equipment was being serviced in Russia, and to explore the possibility of establishing a Ford tractor plant in Russia. Allan Nevins and Frank Ernest Hill describe the result of this visit in their book, *Ford: Expansion and Challenge, 1915–1933*. The Ford men found that the Fordson was far the commonest tractor in Russia, and it was also generally preferred, except in some areas where the extra power of International tractors was favoured.

In spite of poor standards of servicing and the inadequacy of some locally made spare parts, the tractors were performing well, and in some areas were working a night shift with the light of headlamps. The Russians were starting to produce their own copy of the Fordson, called the Krasny Putilovitz. The cost of this tractor was $2,200, compared with $900 for the imported Fordson complete with plough.

The visit of the Ford delegation to Russia eventually resulted in a deal, which contributed to Stalin's first five-year plan. Under the terms of the deal Henry Ford agreed to provide the Russian Government with the technology to build and equip a factory to produce 100,000 cars and trucks a year. Ford saw this as an opportunity to contribute to world peace and prosperity, and his terms appear to have been generous. The financial record for the complete transaction, which included Russian purchases of vehicles and spares from America, indicate that Ford lost about $500,000.

The large numbers of Fordson tractors imported into Russia, plus the copies made at the Putilovitz factory, formed the basis for Russian farm mechanisation in the 1920s. Later, as the Russians developed a liking for tracklaying vehicles, production of what appears to have been a copy of the American Caterpillar tractor became important.

In America, Fordson tractor production continued until 1928, reaching a level of over 100,000 units a year in 1923 and 1925. Between 1928 and 1939, production was concentrated on the factory in Cork, Ireland, but was resumed again at Dearborn, Michigan with the Model 9N. This Ford tractor with Ferguson System brought Henry Ford and Harry Ferguson together in a highly successful but temporary partnership.

4

PROGRESS BETWEEN THE WARS

After the end of the First World War demand for tractor power increased. This situation tempted a number of new manufacturers into the market, particularly in Europe, where the industry had been small-scale and extremely fragmented before the war. Among the manufacturers producing tractors for the post-war boom was the Anglo-Hungarian company, Hofherr-Schrantz-Clayton-Shuttleworth, with a factory in Budapest. This firm was a link between the partnership of Nathaniel Clayton and Joseph Shuttleworth, and an Hungarian syndicate. Clayton and Shuttleworth had established an excellent reputation for their steam engines, produced at a factory in Lincoln, and Hofherr Matyas and Schrantz Janos had a farm machinery business in Budapest. The H.S.C.S. business began in 1900, and the first tractor bearing these initials—the name of the company was presumably too long to appear in full—was tested in 1921 and by 1923 H.S.C.S. were marketing a range of tractors. Later they adopted the diesel engine, which appeared in Britain in 1930 in an H.S.C.S. tractor entered for the R.A.S.E. Trials. In 1955 the name of the company was changed to Dutra, an abbreviated form of the words dumper and tractor.

Another east European country to develop a tractor industry on a commercial scale at this time was Czechoslovakia. Two companies were active in the early 1920s, both with motor ploughs. One was the Praga organisation, better known as a car manufacturer. The first Praga motor ploughs had been produced before the war, using the four-cylinder petrol engine developed for the Praga Grand car. Later models used a specially designed 45-h.p. engine, and were exported in considerable numbers. The other company was Laurin and Klement, also a car manufacturer and the immediate forerunner of the present-day Skoda car and tractor company. Their Excelsior motor ploughs were fitted, at least at one stage, with a driving compartment which looked remarkably like the rear half of a vintage touring car body. This was complete with doors, and a canvas hood on a metal framework which could be folded back in fine weather.

The most active development in Europe occurred in France and Germany. Renault's entry into the market via their military light tank has already been described. Peugeot also appeared after the war with a tracklaying tractor, and Citroen produced a particularly neat looking lightweight tractor especially for vineyard work. At the other end of the scale was the 35–40-h.p. Lefebure tracklaying tractor weighing more than 10,000 lb.

A French company, Latil, failed to achieve the success it apparently deserved. Towards the end of the war they were making a three-wheeler tractor, with a single rear driving wheel which must have been quite unstable. However, their Model TLA, produced in the late 1920s, was outstanding for its advanced specification. This tractor was powered by a 17·9-h.p. engine with four cylinders and a three-bearing crankshaft. Six forward and two reverse speeds were provided through the use of a second gear box providing high and low ratios. The driver, positioned between the axles and provided with a seat with back and arm rests, had both a hand brake and a foot brake. The tractor was equipped with four-wheel drive, and a differential lock was provided. The belt pulley was an optional extra, but could be operated at three speeds, and a capstan or winch, also optional, had positive drive for both forward and reverse working. The tractor was on sprung axles, and the drawbar was also sprung to absorb shock loads. Three types of wheel could be used with this tractor: ordinary steel wheels with lugs, disc wheels which could be fitted with high-pressure inflatable Dunlop tyres, or the rubber tyres plus folding extension plates which moved out of the way for roadwork, but provided additional grip in the field.

A development which appeared on several tractors soon after the First World War was a power-operated lift for rear-mounted implements. In a contemporary photograph the French Tourand-Latil 35-h.p. tractor is shown with a five-furrow mounted plough which could be raised or lowered by means of a cable attached to the plough body and passing over a crane-like framework at the rear of the tractor. The design of the Tourand-Latil looks remarkably like the front end and driving compartment of a vintage truck, and the crane at the rear gives the complete unit the appearance of a breakdown vehicle. A British manufacturer, Burnstead and Chandler of Hednesford, Staffordshire, produced a similar design of lift, operated with chains from the plough body to a winch, for their Ideal tractor of 1918. Heider of America had a foot-controlled winch to power-lift implements on their Model D of about the same period.

German tractor manufacturers were the first to see the potential of the heavy oil or diesel engine, and to make a commercial success of it. Lanz, the company which later became the largest tractor manufacturer in Germany and eventually became part of the John Deere organisation, produced their first tractor in 1921. This was a 12-h.p. model of original design, called the Bulldog, which used a single-cylinder semi-diesel or hot-bulb engine. Lanz and several other continental firms, including H.S.C.S. and Mercedes Benz, persisted with this type of engine, which was reliable and simple, and which could burn a variety of fuels. Among the British manufacturers using diesel engines by 1930, were Marshall, with a single-cylinder semi-diesel engine started by blowlamp, and Aveling-Garrett and Blackstone both with four-cylinder high-speed true diesel engines. The Aveling-Garrett (or Aveling and Porter) engine was started by an electric motor, and the Blackstone used an auxiliary petrol engine. The first American manufacturer to fit a diesel engine commercially was Caterpillar in 1931, when they announced the Model 65 with a four-cylinder four-stroke diesel operating at 650 r.p.m. and started by a two-cylinder petrol engine.

Italy's tractor industry also dates from about 1918 when Fiat produced their highly successful Model 702. Landini, now part of the Massey-Ferguson empire, entered the market for the first time in 1924. Francesco Cassani, who later founded the Same company, completed his first prototype tractor in 1927. This was remarkable for its early use of a diesel engine, and also because its designer was then only twenty-one years old.

The period between the wars was a difficult time for Britain's tractor industry. Of the three manufacturers that had played the major role in establishing Britain as the leading exporter of tractors for several years before the First World War two, Saunderson and Ivel, had vanished completely by the early 1920s and the third, Marshall, had temporarily withdrawn from the market. Marshall re-entered the market in 1930 with a diesel tractor, forerunner of a popular series of single-cylinder diesels which the company produced until the mid-1950s.

Of the companies which launched tractors to catch the period of brisk demand in Britain after 1918, few produced machines of real interest or novelty, and few survived the economic depression of the early 1920s. The most successful British tractor of this period in terms of sales was the Austin, a light but conventional machine with a four-cylinder side-valve engine developing 25 h.p. Production of the Austin was discontinued in Britain after 1921, and the British market was supplied by machines imported from France where they were manufactured under a licencing agreement. Blackstone of Stamford, Lincolnshire, produced both wheeled and crawler versions of their 25-h.p. tractor. The engine for this tractor was an interesting design. It had three cylinders and a self-starting mechanism operated by compressed air. While most paraffin or kerosene engines for years to come were designed to start on petrol, the Blackstone could be started on paraffin by advancing the ignition and using fuel injection. Peterbro tractors, manufactured by Peter Brotherhood Ltd. of Peterborough, also used an unconventional engine, in this case designed by Harry Ricardo. There was provision for diverting some of the exhaust gas from the four cylinders through to the carburettor to preheat the intake air. The crosshead engine was also designed to minimise the amount of unburned fuel entering the sump.

It was becoming increasingly obvious that tractor sales were a growing threat to the future of the steam traction and ploughing engine manufacturers in Britain. The news in 1912 that Daimler were producing a tractor with a six-cylinder engine developing 105 h.p. and claimed, incredibly perhaps, to pull twenty-one furrows, was bad news for the men of steam, even though the Daimler project was ended by the war. Even worse was the news, in 1918, of the production volume and low price of the Fordson. Several British steam engineering companies attempted to salvage the situation just before and just after the war by adapting the internal combustion engine for cable ploughing. Fowler of Leeds announced their first cable ploughing tractors in 1912, and were promptly criticised for designing the tractors, with a dummy funnel and boiler, to look like steam engines. Fowler used a horizontal drum to operate the cables pulling their plough, with the drum mounted beneath the 'boiler'. McLaren, another Leeds firm famous for their steam engines, entered the internal combustion double-engine ploughing tackle section of the R.A.S.E. trials in 1920, with engines operating vertical winches mounted at the rear of their ploughing tractors. Walsh and Clark of Guiseley, Yorkshire, sold a number of their Victoria engines for cable ploughing during and just after the war, and some of these have been preserved. The fuel tank of the Victoria was claimed to hold enough paraffin for a week's work.

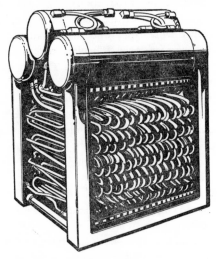

High-pressure tubular boiler used in the Bryan steam tractor.

While some British firms were attempting to use the internal combustion engine to do the work of steam, some American designers were attempting to adapt the steam engine for tractor work. The American approach was to replace the conventional steam boiler, with its disadvantages of large size, heavy weight and high rate of water consumption, by a compact arrangement of tubes which could operate more efficiently and under great pressure. A tubular boiler heated by an oil burner could be as compact as an internal combustion engine, and a steam tractor using a power unit of this type could be similar in external appearance to a conventional tractor. International Harvester produced a prototype steam tractor in 1923, operating at a pressure of 600 lb. per square inch and weighing just over 5,000 lb. The Bryan Harvester Company of Peru, Indiana, began experimenting with high-pressure tubular boilers in 1920, and two years later announced their 15–30 Light Steam-Tractor.

Most of the development effort in America during the 1920s and 1930s was directed at making the internal combustion engined tractor more reliable and more versatile. With a few important exceptions, including the use of diesel engines and the introduction of the Ferguson System, most of the progress in tractor design during this period came from America.

In 1929 John Deere introduced a greatly improved version of the power-operated lift for raising implements, and in 1933 Allis-Chalmers popularised the low-pressure inflatable tractor tyre. International Harvester introduced two of the most important developments just after the war, when they marketed the power-take-off device which had originated from France, and a few years later launched the Farmall tractor. The p-t-o first appeared commercially on the International Harvester 8–16 model of 1918, and it was standard equipment on later versions of the same model. Imitators were quick to copy this, resulting in much confusion over different rotation speeds and directions, until the industry agreed to standardise in 1927.

The Farmall design, which appeared in prototype form in 1922, and which was extensively field tested the following year, was the idea of an International Harvester designer, Bert R. Benjamin. His aim was to adapt the standard tractor design to suit the vast acreage of row-crop cultivations in the American corn belt and elsewhere. He had already experimented with a row-crop machine, which had rear-wheel steering and front-mounted cultivator bodies. International were apparently hesitant about marketing Benjamin's design, but they took the decision in 1925, probably prompted by the havoc which Henry Ford was then causing in the American tractor market. Faced with falling demand, Ford's approach to maintaining his very high volume of production was to slash his prices. The policy, which put the ex-works price of a Fordson below production cost, meant a loss of millions of dollars. Ford was rich enough not to care, and in fact his wealth was so great and his accounting so scanty that he probably had little idea of the extent of his losses. But if Ford was willing to lose money, other manufacturers were not. Many were forced out of business as the Fordson share of the American market climbed towards seventy-five per cent. The Farmall tractor was an answer to the Fordson, and it was first launched in Texas, where it was an immediate success for cotton growing.

The first Farmall was essentially the style of row-crop tractor which is still important in the States. The twin front wheels, mounted close together in a tricycle design, high ground clearance to work over growing crops, and good visibility and manoeuvrability were all featured in Benjamin's design. Implements could be mounted behind, beneath or in front of the Farmall, and differential or steering brakes were fitted for close turning at the end of rows.

Although the most noticeable improvements to tractor design were the introduction of items such as power-take-off, row-crop design and inflatable tyres, equally important progress was being made in details such as improved air cleaners, the use of better materials and manufacturing methods, and improved layout of controls and instruments. During the 1920s chrome-nickel steel was adopted generally for crankshafts, and cylinder liners were being made of nickel iron. Ball and roller bearings were used increasingly to reduce wear and maintenance, and parts likely to wear quickly were more effectively enclosed to prevent entry of dust and soil particles. Air cleaner design became better, and the efficiency of cleaners gradually improved from less than fifty per cent to nearly one hundred per cent.

These improvements helped raise the efficiency and reliability of tractors, reducing the need for routine maintenance. The instruction book for one International model, dating from about 1915, gives some indication of the servicing problems then involved in a day's work. Eleven points on the tractor required grease or oil twice a day, five needed attention every two hours and the rear axle bearings needed greasing by hand every hour. Starting procedures on tractors, in the days before electric starter motors, diesel engines and efficient lubricators, could be an exacting, energetic and time-consuming process. Fuel had to be hand-pumped into the carburettor, mechanical oilers needed priming with perhaps fifty turns of a handle, and the carburettor had to be adjusted for paraffin, and later readjusted for petrol. Engines were cranked either by manually spinning the flywheel or by a cranking handle. Many drivers, faced with a temperamental engine, and badly filtered fuel which blocked fuel lines and needle valves, must have cursed the day when the farm horses were sold to make way for the iron horse.

Rubber tyres

The Allis-Chalmers Model U, introduced in 1929, was basically a conventional and rather uninteresting tractor in its standard form. But it became famous when it was featured in the publicity campaign organised by Allis-Chalmers to advertise the advantages of the low-pressure inflatable tractor tyres which had been developed in 1932.

The theme of the sales campaign was speed, and in 1933 a specially prepared version of the Model U tractor was driven over a mile course at Dallas, Texas, at an average speed of 64·28 m.p.h. The event, which was one of a whole series of publicity stunts, was officially observed by the American Automobile Association, and was recorded as a world speed record for agricultural tractors. The driver for the occasion was Barney Oldfield, who had earlier established an international reputation as a racing car driver. His speed record in the Model U apparently remains unbroken.

For years tractor designers had been trying to devise driving wheels smooth enough to allow tractors to be driven on public roads without damaging the surface, but which would also give sufficient grip to pull a plough on wet soil. The problem had produced a remarkable number of answers, none of which had been entirely satisfactory. While this situation lasted, advocates of horses for farm work could still claim that the tractor lacked the versatility of animal power. A more serious result of failure to solve the problem was the banning of tractors with spade lug wheels from properly surfaced roads, because of the damage they caused.

Tractor manufacturers tried to meet the problem with smooth metal bands or over-tyres. These could be fitted outside the cleats or spade lugs of the driving wheels for road travel, but removed when the tractor was on soil. This idea helped to solve the difficulty, but also had disadvantages.

The time taken to fit the rims to mud-caked wheels was considerable, and when the rims were fitted they made travelling on a tractor over a hard surface extremely uncomfortable. The same objections applied to the obvious alternative, which was to fit detachable spade lugs for field work, and remove them to expose a smooth rim for the road.

The idea of using rubber on tractor wheels was not new when Allis-

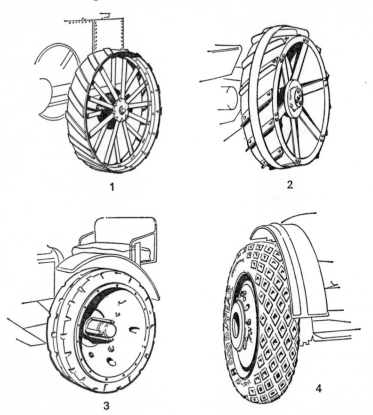

Progress in tractor wheel design. 1, Metal bars attached to the rim, suitable for roadwork, but giving insufficient adhesion on wet land, (Ivel, 1902). 2, Metal lugs, with a detachable steel band for use on roads, (Renault PE, 1920). 3, Solid rubber tyres, (Deutz MTZ 220, 1931). 4, Low pressure inflatable rubber tyres, (Allis-Chalmers U, 1933).

Chalmers were promoting their low pressure tyres. Solid rubber bands had been used on steam traction engines in America and Europe for road travel, and the same idea had been adapted for tractor wheels. The British Simplex tractor, manufactured in Kent, was available in 1919 with a spare set of rubber rimmed wheels as an optional extra priced at £70. High pressure pneumatic tyres were widely used on industrial versions of agricultural tractors during the 1920s, and the French Latil tractor was imported into England in 1929 with high pressure inflatable tyres. In America the Firestone Company experimented with solid rubber blocks set in a metal base for attaching to tractor wheels. These were tested in 1918 on an International tractor, but with unsatisfactory results. The idea of fitting metal studs projecting through the surface of solid metal tyres proved to be an unsuitable compromise. When the studs projected far enough to grip the soil, they were also sticking out too far for road travel.

The breakthrough came with the development of the low pressure tyre. This could mould itself to the contours of a furrow bottom or a hard, stony surface, it could grip as efficiently as steel cleats, and also absorb some of the bumps on a hard farm track so permitting faster speeds without bouncing the driver off his seat.

Allis-Chalmers began their experiments with tyres designed for aircraft. Results were promising, and a Model U tractor, equipped with Firestone tyres inflated to 15 lb. pressure, was tested on a farm near Waukesha, Wisconsin in April 1932. Allis-Chalmers were soon satisfied that they had achieved the long-sought breakthrough. The company was eager to expand their modest share of the American tractor market, and with the co-operation of leading tyre manufacturers they launched the low-pressure tractor tyre towards the end of 1932. The initial response was disappointing. Farmers were by no means convinced of the advantages of rubber tyres, and there was some concern about the likelihood of punctures.

In 1933 the company launched a massive and ingenious publicity campaign. They stressed the increased speed which their tyres permitted, rather than the technical advantages of improved versatility or greater efficiency in the field. The campaign started with a team of suitably modified tractors being taken on a tour of the State Fairs where they were raced at speeds of 30 m.p.h. or more. Well-known racing drivers were hired to drive the tractors, and the races became a popular spectacle.

47

Another publicity stunt in the same year involved a trip by a Model U tractor from Milwaukee, the Allis-Chalmers home town, to Chicago. The event, timed to attract the attention of farmers attending the International Livestock Exposition in Chicago, involved driving eighty-eight miles on the highway, a distance covered in five hours and one minute. Some Allis-Chalmers dealers are reported to have driven tractors deliberately at speeds well above the maximum permitted on public roads, in order to gain convictions for speeding. The publicity caused by getting a ticket for driving a tractor too fast could mean extra sales. The world speed record established by Barney Oldfield in Dallas was the last in a whole series of record attempts.

The publicity campaign was successful. Allis-Chalmers increased their sales and their share of the market, and other tractor makers in the United States, and more gradually in Europe, were forced to follow their lead and fit similar tyres. Behind all the stunts and the noise of the highly tuned engines, Allis-Chalmers had made a genuine contribution to tractor development. Their energetic sales campaign encouraged the rapid adoption of tyres which made tractors more versatile, more efficient and less fatiguing to drive.

The Model U, which had demonstrated the advantages of low pressure tyres, was in other respects only a modest success. In 1934 it was replaced by the WC, claimed to be the first tractor designed specifically for the new tyres. This tractor was priced at $825 with tyres, and a steel wheel version cost $150 less. By 1937 almost half of the new tractors sold in the U.S. were fitted with low-pressure inflatable tyres.

The Ferguson System

The efficiency of modern farm tractor design owes much to Harry Ferguson. The system he developed for attaching implements directly on to the rear of a tractor, with weight transfer and automatic draft control, is now used in some form by most of the world's tractor manufacturers.

Harry Ferguson was born in 1884 in what is now Northern Ireland. His father, who had eleven children, farmed a hundred acres near Belfast. Time spent at home on the family farm may have provided Harry Ferguson with experience which would be helpful later in his career, but there is little evidence at that stage that he cared much for farming. He appears to have welcomed the opportunity to leave home when he was eighteen to join an elder brother in a cycle and car repair business in Belfast. He quickly showed a talent for mechanical engineering, and particularly for coping with the temperamental engines of the early 1900s. He also took an active interest in motor cycle and car racing, and later played a major role in establishing the famous Ulster TT motor races which attracted some of the world's leading drivers to the Ards circuit. Another early outlet for Harry Ferguson's engineering skill was aviation. In 1908, when aircraft were still a novelty and few people realised that they might have commercial or military value, Ferguson began designing and building his own monoplane. He flew it successfully for the first time in 1909.

During the First World War, Harry Ferguson, by now running his own garage business, took up the agency for the Overtime tractor, an American design which achieved a fair degree of success in Britain. He took a close personal interest in selling these tractors, and arranged demonstrations with the extraordinary attention to detail which later marked his attempts to sell tractors bearing his own name. As the war progressed and the U-boat campaign brought Britain close to starvation, the Government recognised the need to make maximum use of the few tractors available. The Irish Board of Agriculture asked Harry Ferguson to make a tour of Ireland to encourage more efficient utilisation of tractors. Ferguson, and his assistant engineer, William Sands, visited farms and organised numerous demonstrations during 1917. His efforts to coax satisfactory results from cumbersome tractors dragging implements which were designed for horses to pull, encouraged Ferguson to consider more efficient ways of achieving the same result.

When he returned from his demonstration tour, Ferguson and his assistants began work on an entirely new plough design. This was completed in 1917 and was effectively the beginning of the Ferguson System. The plough, designed for a relatively light tractor, was intended to be attached directly to the tractor. The plough could be lifted out of work by means of a lever operated manually from the tractor seat, and was coupled close behind the tractor to give more constant working depth. The most important feature of the design was the geometry of the linkage between the tractor and the plough, which utilised the resistance of the soil against the plough to provide a downward force on the tractor wheels for extra grip.

The first Ferguson plough was designed to fit on to the Eros tractor. This was one of several conversions available for the Ford Model T car. For field work the rear wheels of the car were replaced by ones with larger diameter, and the engine and gearing were modified to allow prolonged operation at high engine speed and slow forward speeds. The Eros was popular, and the Ferguson plough might have achieved a worthwhile volume of sales. However, in 1917 tractor conversions for the Model T were already doomed on both sides of the Atlantic by the arrival of the Fordson tractor.

Harry Ferguson reacted quickly to the success of the Fordson. He designed a new plough, based on his original ideas but with some improvements, for attachment to the Fordson. In 1925 he formed a company in the States to manufacture and market the plough and the success of this venture provided finance for further development work. One of the advantages of the Ferguson plough was that it helped overcome the alarming tendency of the Fordson to rear over backwards. The light front of the Fordson lifted all too easily when a trailed implement, conventionally hitched, came up against a solid obstruction. With the forward movement of the implement suddenly stopped, the energy in the engine and flywheel immediately became a lifting force near the front of the tractor. This fault was not unique to the Fordson, but the relatively short wheelbase and light weight of the tractor tended to accentuate the risk, and the driver, perched at the rear, could be in serious danger. The hitch arrangement of the Ferguson plough tended to combat this risk. The design of the parallel struts was intended to transfer forces from a stalled implement through to the front of the tractor, and gave a degree of rigidity to the hitch.

The next stages of development in what later became the Ferguson System, were the use of hydraulic control over implement depth, and a form of three-point linkage. This opened the way to the incorporation of automatic draft control. When a plough or tined cultivator moved into harder soil, the extra resistance could be sensed mechanically and cause the implement to be raised slightly. This produced a two-fold effect: the

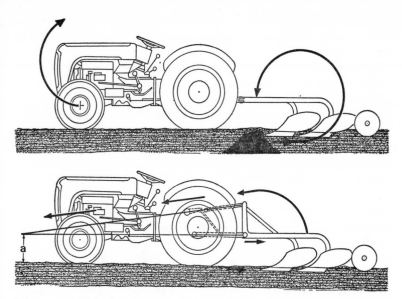

Early tractors, with implements attached to a low drawbar, were liable to overturn backwards when the implement met an obstruction. The geometry of Harry Ferguson's attachment converted resistance against an implement into a force which helped to stabilise the tractor.

reduced depth of work helped the tractor and implement through the area of difficult soil, and more of the implement's weight was put on the tractor wheels to give improved adhesion.

With his ideas basically completed, Harry Ferguson began to look for the capital and the manufacturing capacity which could put the system on to the market. A major obstacle was the fact that the tractor and the implements had to be designed to suit the system, and this made the project a substantial operation. Another factor which probably deterred some potential manufacturers, was the prospect of having to work in close association with someone as dogmatic and self-assured as Ferguson. After being turned down by several companies, Harry Ferguson decided to build a prototype tractor incorporating his ideas, which he could use to demonstrate their value. The result, a compact, lightweight tractor powered by an American Hercules engine of 18 h.p., was completed in 1933.

This tractor, painted black and known as the Black Tractor, introduced one of the greatest developments in farm mechanisation, and is displayed at the Science Museum in London.

The David Brown Company of Huddersfield, Yorkshire, then a family controlled firm specialising in gear manufacturing, became the first company to build tractors incorporating the Ferguson System. The grey painted Ferguson-Brown tractor was launched publicly in May, 1936. The first version of this tractor was powered by a British Coventry Climax engine developing 18–20 h.p., but this was later replaced by a 20-b.h.p. David Brown engine. The tractor had three forward gears and one reverse and was equipped with independent wheel brakes. A particular feature of the design was that nuts and bolts were standardised to only two sizes, so that one double-ended spanner could be used for all adjustments in the field. The spanner was also marked off in inches so that it could double as a measure. Harry Ferguson retained full control of the marketing, and personally supervised many of the demonstrations, which were always carefully arranged with great attention to detail. The marketing programme included a training school to help machinery dealers and tractor drivers to use the equipment correctly.

In spite of the energetic salesmanship and the genuine advantages of the equipment, sales were disappointing. This was partly because many farmers remained doubtful of the virtues of such an unconventional, small tractor, and partly because of the cost of changing to the Ferguson System. The tractor cost £224, which was about fifty per cent more than the Fordson which then dominated the market. In addition, farmers buying a Fordson could generally still use their existing machinery to work with it, but to buy a Ferguson-Brown meant discarding some conventional machinery. Less than two thousand of the tractors were sold before the partnership broke up in 1939.

The partnership ended after Ferguson took one of their tractors to America to gain the interest of Henry Ford, and David Brown decided to design and manufacture a new tractor with the extra power which farmers appeared to demand. The new David Brown tractor, the VAK 1, was rated at 35 b.h.p. and had the advantage of four forward speeds. The design included hydraulic lift, but not the Ferguson patented system of depth control. This was the first in a series of David Brown tractors which achieved outstanding commercial success. The tractor interests of the David Brown Corporation were taken over in 1972 to join Case as part of the Tenneco organisation of America.

David Brown's VAK 1 was launched in July, 1939, one month after the first of the Ford 9N tractors was released in the United States. The 9N, or Ford Tractor with Ferguson System, was the result of the famous 'handshake' agreement between Harry Ferguson and Henry Ford. The two men had met in the autumn of 1938 and, simply by shaking hands, sealed an unwritten and unwitnessed agreement to join forces to develop and produce a new tractor. As a direct result of this unconventional agreement more than 300,000 9N tractors were built between 1939 and 1947. The new tractor, produced by two of the greatest men in the history of tractor development, was designed around an improved version of the Ferguson System which had first appeared on the Black Tractor. It was marketed in America at a price of $585, which was very much in line with Henry Ford's policy of setting a price level for his tractors which his competitors found almost impossible to match. Harry Ferguson retained full control over marketing, forming a sales organisation for the tractors which Ford manufactured.

The famous handshake agreement ended in 1947, and was followed by an even more famous lawsuit in which Harry Ferguson's company sued the Ford organisation for more than $250 million. Ford had introduced a new tractor, the 8N, which was an up-dated version of the 9N with a number of improvements including a four-speed gearbox. The 8N, which was sold through a marketing organisation formed by Ford, included the Ferguson System—an alleged infringement of some of Harry Ferguson's patents. The enormous value of the claim made by Ferguson against Ford was based mainly on the loss of business involved when the Ferguson sales organisation was by-passed, and also on the alleged infringement of patents. The case was complicated and protracted, and the difficulties in settling it were aggravated by the nature of the original agreement between the two sides. It was hard to determine exactly on what basis the two men had agreed to operate. The case lasted until 1952, when Ferguson rather suddenly agreed to settle for $9,250,000 from the Ford organisation, plus an arrangement whereby Ford ceased to use some Ferguson patented features in the 8N.

Ferguson probably gained some publicity by appearing in the role of a David fighting for the right to commercial survival against a Goliath trying to squeeze him out of existence. Although the Ferguson organisation had obviously suffered through suddenly having nothing to sell in America, by the time the case was settled there had been a remarkable recovery. It is worth noting too, that the Ford attitude to the use of Ferguson patents was probably influenced by the fact that Henry Ford had often allowed other people free use of his own patents.

The tractor which enabled Harry Ferguson to stage his comeback was the little TE–20, manufactured by the Standard Motor Company in Coventry, and also at the Ferguson Park Plant in Detroit where it was known as the TO–20. Production at Coventry began in 1946, before the agreement with Ford had ended. This was possible because the British end of the Ford organisation had turned down the idea of producing a Ferguson type tractor. The design of the 'Fergie', as the TE series was generally called, owed much to the Ford 9N, but included a four-speed gearbox. The engine used initially was an overhead-valve design from the American Continental company. Sales increased rapidly, helped by a substantial export effort, and production at the Coventry factory reached about 70,000 tractors a year in the peak period of 1951–52. In the same years assembly at Detroit exceeded 30,000 tractors a year. The TO–30 model, introduced in the States in 1950, featured slightly more power plus improved controls and a better seat. This model was intended to maintain sales until the anticipated arrival of a new, more powerful tractor.

Before this new model appeared, however, it was announced in 1953 that the Ferguson companies were to be merged with the Massey-Harris organisation to form Massey-Harris-Ferguson Limited. Whether it was a merger, or a takeover by Massey-Harris is debatable, but it was in many ways a logical arrangement. Massey-Harris had achieved outstanding success with farm machinery, but had rarely matched other major full-line companies in the tractor market. By purchasing the Ferguson companies and patents, Massey-Harris acquired the right to use one of the best-known names in the tractor market, the tractors which accounted for well over fifty per cent of British sales at that time, and a strong marketing organisation in many countries. Harry Ferguson, seventy years old at the time of the merger, resigned from the new organisation twelve months after its formation, and effectively retired from active involvement in the world tractor scene.

The negotiations with the Massey-Harris directors provided Harry Ferguson with another opportunity to demonstrate his unconventional approach to big business. Ferguson was travelling by car with three senior directors of Massey-Harris, discussing the difference of $1 million between what he had been offered for his companies and what he was asking. Ferguson offered to settle the matter by tossing a coin. The gamble took place in the village of Broadway, Worcestershire. Ferguson lost his million dollars, but promptly asked for the coin to be tossed again for the right to keep the coin itself. This time he won.

Harry Ferguson retained some active interest in tractor development after leaving Massey-Harris-Ferguson (now Massey-Ferguson). But, in the final years before his death in 1960, he achieved more with inventions for the motor car. These included the Ferguson Formula for improved adhesion and control, which has been used commercially.

5

TRACKLAYING TRACTORS

The theory of mounting a vehicle on crawler tracks instead of on wheels has been known for two centuries, and possibly even longer. Several patents on the idea were established in Britain from the second half of the eighteenth century onwards, but there was little practical interest in the idea until it had been developed and improved in America.

Some attempts were made in Britain to manufacture and market crawler or tracklaying steam traction engines, but the commercial results were disappointing. In America there were situations where the crawler had worthwhile advantages over wheeled vehicles, and it was in these special-ised markets that the idea first had commercial success. The Phoenix Com-pany of Eau Claire, Wisconsin, made steam-powered crawler tractors for forestry work from 1904. In California large areas of fertile but soft peat lands provided difficult working conditions for steam engines with wheels, and the Holt and Best companies both produced crawler tractors for these areas.

Benjamin Holt, whose companies dated back to about 1869, built his first steam-powered crawler tractor in 1904 at Stockton, California. The machine proved its value locally, and Holt was able to develop and es-tablish markets elsewhere in the United States and Canada. In 1908 Holt bought the business, though apparently not the factory, of their rival, Daniel Best. Two years later C. L. Best, one of Daniel Best's sons, started his own crawler tractor business in his father's old factory at San Leandro, California. His Tracklayer tractors were strong competitors to the Holt models, and the Holt company merged with the C. L. Best Tractor Com-pany in 1925. The result of this merger was the Caterpillar Tractor Company of Peoria, Illinois. The name 'Caterpillar' was the Holt trade mark, but the tractors produced after the merger owed much to C. L. Best, who had developed a rigid frame design in place of Holt's semi-articulated frame. In 1931 the Caterpillar Company became the first American tractor manufacturer to market a diesel-powered model with the four-cylinder 65. It became the market leader among crawler tractor manufacturers and developed substantial interests in earthmoving and other civil engineering equipment.

There is a widely held belief that Richard Hornsby & Sons of Grantham, Lincolnshire, developed the crawler principle, and sold their rights to the idea to Holt of California. Britain was then obliged to buy the idea back from Holt when the First World War tank was developed. This theory is unlikely to be based on fact. Certainly Britain's first tanks were based on Caterpillar-type tracks, but these tracks were a great improvement on the Roberts track which Hornsby had used, and Holt had developed his own tracks some time before the sale of patents by Hornsby is supposed to have occurred.

Another American company which achieved success in the crawler tractor market was the Cleveland Tractor Company. This was founded by the White family, manufacturers of the White steam car, who had also developed a motor plough. The Cleveland Company was formed in 1916 and produced the Cletrac crawler tractor, a name which survived commercially for several years after the manufacturing company had been taken over in 1944 by the Oliver Corporation.

The basic design for crawler tractors was established by Holt and Best, and this has since become accepted as the conventional layout. Inevitably there were attempts to introduce new ideas, particularly in America, and some of these were extremely curious in appearance. The Bates Steel Mule, manufactured at Joliet, Illinois, and imported into Britain by the Vulcan Car Agency, was the name given to a series of crawler and half-track machines, of which the Model C was the most memorable design. This was a tricycle layout designed for row-crop work. At the front two wheels steered the machine, and at the back was a single crawler track, centrally positioned, which transmitted the drive from a chain and sprockets. The Model C was powered by a four-cylinder E.R.D. engine, and was produced for about three years until 1918. A feature of the design was the extending steering column which allowed two driving positions, either conventionally at the rear of the tractor or, with the steering column extended, the driver could operate from the seat of the implement following the tractor.

The Killen-Strait was another unconventional crawler tractor produced at much the same time as the Steel Mule C and was also handled in Britain by Vulcan. This, too, was a tricycle-type model, but with three tracks. The leading track was for steering and the two rear tracks transmitted the drive. The layout in this case was not for row-crop work, but for ploughing, and the front track was positioned off-centre, in line with the right hand rear track. The design of the Killen-Strait tracks is believed to have been the fore-runner of the track arrangement of the first tanks (see page 30).

Although crawler tractors have achieved considerable success for civil engineering work, their use in agriculture has been more limited. The development of rubber tyres gave wheeled tractors more versatility and higher working speeds, and the more recent success of four-wheel drive tractors has also tended to limit sales. With a few notable exceptions, including the Renault of the early 1920s, and more recent Carraro, Fiat and Fowler-Marshall crawlers, European manufacturers have concentrated mainly on wheels and left tracks to American and Russian companies. There was a considerable, but temporary, upsurge in the production of crawler tractors in the 1950s in Britain, when eleven companies were competing for the very limited market. These included Fowler, with their Challenger range, the successful David Brown TD series, and Ransomes who were making one of their rare excursions into the tractor market, this time with the tiny MG6 crawler for market garden work. In addition to the eleven British manufacturers, there were also half-track crawler conversion kits available for some wheeled tractors.

Self-propelled implements

The history of self-propelled implements, like that of crawler tractors, dates back to the period when steam power provided the only practical means of propulsion. One of the earliest attempts to incorporate both power unit and implement in a single, self-propelled machine, was the steam plough patented in 1849 by James Usher of Edinburgh.

Usher's machine was a rotary plough, with three sets of five plough bodies mounted on a horizontal shaft behind the boiler. The plough unit was driven by a system of gears from the steam engine, and rotated in the direction of travel. It was claimed that the rotary action of the plough

Usher's Steam Plough, 1849. Drawing based on a model of the original machine, displayed at the Science Museum, London.

bodies tended to push the whole machine forward. The Usher machine, together with some other nineteenth-century attempts to devise self-propelled rotary cultivators, never achieved commercial success. It is possible, however, that these Victorian machines were simply ahead of their time and limited by the poor power-to-weight characteristics of the steam engines then available. One advantage of the modern, tractor-operated rotary cultivator is that power is applied through the implement to reduce soil damage caused by wheelspin in wet conditions.

Another self-propelled machine which was almost certainly ahead of its time was the Self-Moving Potato Harvester, announced in 1904 by Moultons of Chatteris, Cambridgeshire. This machine was operated and propelled by a 20-h.p. Simplex steam engine, and was apparently designed to dig, pick-up and load potatoes into either baskets or a trailer—thus pre-dating much more recent interest in bulk handling potatoes. The machine worked on a single row, and was propelled by a chain drive to the rear wheels.

The Moulton potato harvester appeared at a time when farm labour was plentiful and relatively inexpensive in Britain, and economising on man-power costs was a low priority. If present-day economics had applied when the Moulton machine was invented, there might have been more interest and incentive to encourage further development.

With the availability of internal combustion engines, interest in self-propelled implement design increased and resulted in more practical equipment. The Sharp tractor mower appeared in prototype form in Britain in 1904 with a two-cylinder Daimler engine to provide power for propulsion and to operate the mid-mounted cutter bar. A later version used a Humber four-cylinder engine, and was designed to work as a conventional tractor with trailed implements or as a power-driven mower. The Sharp used stub axle steering before most other early tractors.

In the 1920s several European manufacturers entered the market for the first time with power-driven mowers. These included the Swiss company, Aebi, which experimented with various designs, including a three-wheel machine with a twin-cylinder petrol engine, which was sold in small numbers from 1929. This model had two large driving-wheels mounted in line on the same side of the machine—one at the front and one at the rear—both of which steered. The third wheel was on the opposite side and simply kept the machine upright. The first Fendt tractors were pro-duced in Germany in 1928, and one of the two prototypes produced in that year was a small, self-propelled power-driven cutter bar mower, operated by a single-cylinder diesel engine. Kramer of Germany also started with a lightweight grass cutter, powered by a single-cylinder petrol engine, first produced as a prototype in 1925.

The most successful self-propelled machines were motor ploughs, and from 1915, or thereabouts, these were produced in large numbers in Europe and in the United States. The Deutz two-way motor ploughs produced in 1907, with four-furrow ploughs mounted at the front and the rear, have been described in Chapter 1. The general trend of design ten years later was for much lighter machines, with two large driving wheels

60

and the engine at the front, the plough unit slung underneath in the middle, and the driver perched at the rear. The best known, and probably also the most successful of this type of motor plough, were the American Moline and the British Crawley. Both machines sold in considerable numbers and were exported. These two machines also had advanced features. The Moline introduced electric self-starters commercially in 1917. The Crawley machines were known by the trade name Agrimotor. Both the Moline and the Crawley were powered by four-cylinder engines, and the Crawley could be converted with a kit to work with draft implements like a more conventional tractor.

Some of the German and French motor ploughs were more ambitious but less practical. Hanomag marketed an 80-h.p. two-wheel motor plough in 1912, probably the largest machine of its type produced. In France, Delahaye produced a two-way plough powered by a 32-h.p. engine, and with a four-furrow plough mounted at each end of the machine.

The idea of applying power directly to tillage, rather than through a drawbar, has had considerable appeal since the Usher rotary plough of 1849. A number of designers have attempted to produce self-propelled rotary cultivators, with mixed results. One of these, the French Somua, had two large driving wheels at the rear, with the driver positioned between them. At the front was a single small wheel for steering, with a very long tiller reaching back to the driver, giving a strong impression of an old-fashioned bath chair.

An Australian farmer's son, Arthur 'Cliff' Howard, was the first person to make a commercial success of powered rotary tillage, and his early machines were self-propelled. The prototype rotary hoe, produced in 1920, was powered by a 60-h.p. American Buda engine. The engine operated both the driving wheels and the rotor, which on the first machine was fifteen feet wide and made in five equal sections. It was capable of cultivating 3·5 acres an hour.

The first rotary cultivator produced for sale was made by Howard in 1922, the year he formed his company, Austral Auto Cultivators, then based at Moss Vale, New South Wales. This machine was marketed as a rotary cultivator only, but the basic design had been planned as a complete system of mechanisation for cereal growing. The rotary cultivator unit could be removed and a seed drill or a harvesting unit attached in its place.

61

With the harvester unit attached, the machine was designed to travel in reverse. The most successful self-propelled rotary cultivator was the DH22 model produced in the late 1920s. Later, the general adoption of the tractor-mounted Howard Rotavator replaced the self-propelled units, apart from the hand-controlled horticultural Rotavators. A. C. Howard died in England in 1971, where his company, by then operating on a world-wide scale, is based.

Rotary cultivation generally involves blades or spikes attached to a horizontal shaft, but the shafts can also be vertical, as on the modern rotary ploughs from Germany and Eastern Europe. Fowlers, who were looking for a replacement for their declining steam traction engine business in the 1920s, produced substantial numbers of the Gyrotiller, a self-propelled unit of enormous proportions, with two rotary units at the rear.

The principle of the Gyrotiller was developed by N. C. Storey during the First World War, while he was managing a large estate in Puerto Rico. The machine was particularly suitable for the tough conditions of sugar cane cultivation, and Fowler, hearing about the prototype unit bought the patent rights to it. The first Fowler Gyrotiller was manufactured in Leeds in 1927 and was sold to a sugar estate in Cuba in the next year. This machine, and the next three to be produced, took their power from Ricardo petrol engines which developed 225 b.h.p. The two rotary cultivating units on the rear of the machine gave a total tillage width of ten feet. The cultivating tines were capable of revolving at a depth of up to twenty-two inches, when they absorbed maximum power resulting in petrol consumption of up to fourteen gallons an hour. From the fifth Gyrotiller onwards, Fowlers used a more modest, and less thirsty, diesel engine from M.A.N. of Germany, which developed a mere 137 b.h.p.

Most of the machines built were exported, especially to sugar estates in the West Indies. The first British customer was an Essex farmer who bought a Gyrotiller for working his heavy land. The manufacturers launched a sales drive in Britain, following this first home sale in 1932, and a worthwhile market was developed, especially among contractors. In 1939, as industry switched to production of war equipment, production ceased and was never resumed. At the time of writing several Gyrotillers are still being used by contractors.

6

RECENT PROGRESS

Some of the affects of the First World War on the tractor industry were repeated after the Second World War, but to a less pronounced extent. Demand for tractor power increased, as reflected by the production figures in America and Europe in the period after 1945, and there was, too, a marked, but temporary, increase in the number of tractor manufacturers.

The most significant trends since 1945 have been a steady improvement in tractor design, increasing attention to driver comfort and safety, greater interest in four-wheel drive, more power, plus some notable attempts to improve upon the conventional layout of the wheeled tractor.

The improvement in design has been mainly a process of detailed modifications adding up to greatly improved standards of reliability and versatility. Diesel engines, already well established in Europe before the war, have rapidly replaced petrol and paraffin to become virtually the standard power unit in most areas of the world. Electric starting and adequate lighting for nightwork have also become standard equipment, along with rapid acceptance of power steering. Several manufacturers have provided power-operated means of adjusting the track widths of rear wheels. Engine outputs have been raised by the use of turbochargers.

A major step forward was the introduction in 1947 of independent power take-off. This development came from a comparatively small Canadian manufacturer, the Cockshutt Plow Company of Brantford, Ontario, now part of the White Farm Equipment Company of America. This made it possible, for the first time, to stop or change the forward speed of a tractor without affecting the operation of p-t-o operated equipment. The advantages of this system were obvious, especially with the use of machines such as the Allis-Chalmers Rotobaler. Other manufacturers were quick to follow the lead of Cockshutt, and these included Allis-Chalmers who marketed their version on the WD tractor in 1948.

Tractor transmissions have been developed considerably since 1945, when few manufacturers of wheeled tractors offered more than three or four gear ratios. Some field operations demand very precise selection of forward speed, and farmers have proved their willingness to pay for the complexity of twelve or more forward speeds, usually provided by the use

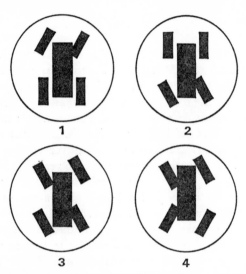

Four-wheel steering, an alternative to articulated tractor design to give extra manoeuvrability to four-wheel drive tractors.

of a second high/low ratio gear box. Synchromesh on some ratios, and a few cases on all, is more widely available for easier, quieter gear changing on the move, and hydrostatic transmission for even more precise selection of forward speed has been introduced, notably by Lely, International Harvester and by several manufacturers of self-propelled equipment, including combine and forage harvesters.

Although the general trend of the industry in America and in Europe has been to concentrate tractor manufacturing into a smaller number of

64

giant international companies, there is still a market for the smaller, specialised company. This situation is similar to that in the motor industry, where smaller companies can survive by concentrating on the sectors of the market which the giant corporations, relying on high volume sales, tend to neglect. In the car industry the specialist markets are for luxury or high-performance vehicles. The tractor industry offers scope for limited sales of relatively expensive four-wheel drive equipment, usually with more power than the major manufacturers offer. Steiger and Versatile in North America and County and Roadless in Britain are some of the companies which successfully meet this demand.

One of the outstanding features of the period since 1945 has been the amount of innovation as some manufacturers have attempted to break away from the conventional layout of wheeled tractors. Much of this new thinking has been concerned with improving the operator's area of vision, and there have been several attempts to re-introduce forward-control layout. Both the County Forward Control model from Britain and the more recently introduced Deutz Intrac system from Germany, have the driver's cab mounted over the front axle. This gives excellent forward visibility, and the Deutz takes advantage of the forward layout by providing three-point linkage at the front as well as at the rear. A secondary advantage of the forward control arrangement is the space which this leaves behind the cab. Both County and Deutz have used this to provide a rear platform on the tractor for mounting equipment such as chemical spray tanks, with loss of visibility for working with trailed equipment.

The idea of providing implement mounting points at the front as well as at the rear, as on the Intrac, appears to be too logical to be ignored. Several other continental manufacturers have incorporated front end hitch points, including Daimler-Benz who announced their MB Trac 65–70 model in 1972, the same year as the announcement of the Intrac. The Swiss company, Schilter, announced front and rear implement attachment for their UT range of tractors in 1973. These, like the MB Trac models, are four-wheel drive, but with the cab mounted between the axles. The MB tractor has a load-carrying platform over the rear axle, the Schilter can carry spray tanks or other loads on its cab roof. With increasing interest in the advantages of performing multiple operations in a single pass, demand for fore and aft implement attachment points seems likely to increase.

Another feature of the new configuration tractors from Europe which should become increasingly popular is the provision of a mounting platform for load carrying. This is not a new idea. In Germany both Fendt,

The Schilter UT72000 from Switzerland, with attachment for front-mounted implements on a loader-type boom, and front p-t-o. Powered implements can also be attached conventionally on the rear three-point linkage. This tractor also has 4-wheel steering.

with their One-man System, and Daimler-Benz with the Unimog have provided tractors which can operate equipment and carry supplies for the job simultaneously. This means that when seed drilling, spreading fertiliser or applying a chemical spray, the tractor can carry seed or chemical, thus avoiding some of the time lost travelling to and from a storage point to refill.

The Fendt One-man System is one of the most successful attempts so far to design a tool-carrying tractor with excellent visibility for row-crop work. With the front-mounted load-carrying platform removed, the

66

driver has an almost uninterrupted view of the ground behind the front axle. This is ideal for working with mid-mounted implements for operations where precise steering is necessary. The Fendt system, with its apparent advantages, has remained essentially European, and it has attracted little interest in other areas. The simple, lightweight tool-carrier principle has been introduced in America and in Britain with little success. Allis-Chalmers launched their Model G tractor in 1948, with a design which included a rear-mounted engine, mid-mounted driving seat and a simple tubular steel frame to act as the tool-carrying attachment for mid-mounted implements. The David Brown 2D, produced between 1955 and 1961, was a similar configuration but the tubular frame of the 2D acted as the reservoir for the compressed air system which operated the hydraulics on this model. Neither of these two tool-carriers attracted a large volume of sales.

The Deutz and the Fendt equipment already referred to are both attempts to introduce new concepts of mechanisation systems based around a single, versatile power unit. There have been other attempts to develop complete systems in this way, starting with the Moline Universal and continuing through the Ferguson System, with its range of implements matched to the tractor.

One of the most notable attempts to produce a power unit as part of a specially devised mechanisation system was the American Uni-Farmer, introduced in 1950 by Minneapolis-Moline, but later taken over by New Idea and marketed under the Uni-System trade name. The central item of equipment in this range is the power unit, available with three engine sizes at the time of writing, and with large diameter driving wheels at the front and small wheels at the rear. The operator's cab is located right at the front and to one side. This gives good forward-control vision and leaves room for equipment or a load-carrying platform at the other side. With the Uni tractor there is a range of harvesting equipment specifically designed for attachment to the power unit. This includes corn or maize harvesters, combine harvesters and green-crop harvesting equipment. The Uni-System also includes other attachments, including snow blowers and a tool-carrier frame for front mounting. At the time of writing the lack of tillage machinery to complete the Uni-System line appears to be a disadvantage. This system is one of the few attempts by a major American tractor manufacturer to break away from conventional configuration since the war, and it is significant that Minneapolis-Moline later sold the system to a smaller implement company then keen to enter the tractor market. Uni-System has been a success, but after more than twenty years on the market it had failed, at the time of writing, to cause much of a revolution in farm mechanisation.

New ideas in tractor configuration are tending to come from Europe, rather than from the giant American or Russian factories. This is partly because the small scale of farming in some areas of Europe demands the versatility of systems such as the Fendt or the Swiss Aebi, and partly because the great international companies have less incentive to gamble on the commercial acceptability of unconventional ideas. To some extent this approach has probably tended to stifle new thinking. The majority of farmers accept that the conventional tractor is satisfactory and matches their existing requirements. Many of the innovations in design which have appeared from time to time with loud publicity, have proved to be expensive failures, or to have limited application. Introducing radical change to tractor design is a gamble, but fortunately there are plenty of manufacturers prepared to take a risk. Progress in farm mechanisation has been caused by new ideas which were often only accepted slowly and with hesitation.

The Driver

The increasing emphasis on safety and comfort is a welcome development. It follows more than half a century of concentrating almost entirely on developing greater reliability, versatility and work output, with the driver being taken very much for granted. During this period, tractor designers did little more for the driver than to provide him with a seat, and eliminate some of the more obviously lethal features which had appeared in the earliest tractors.

Some of the very early tractors must have been extremely dangerous to operate. Several models produced up to about 1915, particularly in America, provided no seat for the driver. He had to stand to operate the controls, and the floor would inevitably become slippery in wet conditions. If he slipped, or if the tractor lurched and threw him off balance, the risk of death or serious injury was high. There was no barrier to prevent the driver falling backwards into the path of the implement being pulled behind. To fall sideways or forwards was equally dangerous because of the unguarded moving parts of the tractor within easy reach. These moving parts included the spoked driving wheels with steel lugs, gear wheels, chains and sprockets, and sometimes even the flywheel, weighing hundreds of pounds.

The situation was made worse by the awkward operation and location of many of the controls, and by the lack of effective brakes. Drivers were generally novices, and although some had gained previous experience with steam traction engines, most had worked only with horses or mules before taking the controls of a tractor.

Not all of the first tractors were equally dangerous, and some of the more obvious hazards were eliminated quite rapidly as tractors became increasingly used for drawbar work rather than stationary operation. However, in spite of the progress made, there were still plenty of tractors in production by the end of the First World War, which had completely exposed driving wheels or transmission chains moving within a foot or so of the driver.

Men who worked in such dangerous situations may not have been concerned about weather protection, but canopies to keep the rain off appeared as either standard equipment or an optional extra at an early stage of tractor development. These canopies were usually almost as long as the tractors and protected the engine as well as the driver. This was useful because early types of ignition equipment became even more temperamental in wet conditions. In fact it was probably considered more important to keep the engine dry than to protect the driver. Indeed some tractors, including the 1892 Case prototype, had canopies extending over the engine but apparently leaving the driver exposed. Canopies, which had been inherited from steam traction engines, gradually disappeared as tractor manufacturers found it cheaper and more practical to put a cover or hood directly over the engine. This, however, left the driver completely exposed to the weather.

The driver's seat is a good example of the attention designers have paid to comfort. The bare metal pan seat, mounted on a spring steel support, first appeared on horse-drawn equipment, and became standard equipment for tractors. The Ivel tractor of 1902 used this type of seat, and it was still in use on the Ferguson fifty years later, as it had been on almost every make of tractor produced in quantity in the intervening period.

The metal pan seat, which was actually more comfortable than it looked, was often padded by the driver with an old sack. The idea that the tractor manufacturer should provide the padding became generally accepted in the early 1950s, although padded seats had become standard equipment on some crawler tractors almost twenty years previously. One of the first really comfortable tractor seats was the driving seat of the Model T Ford car with one of the tractor conversion kits fitted. These were sold from about 1916.

Safety and comfort are now major priorities in tractor development. Cab design has become a sales point, and some manufacturers give as much space in their sales leaflets to describing the tractor seat as they do for the engine or the hydraulics.

If manufacturers had paid as much attention to safety many years ago, lives could certainly have been saved. We have had to wait so long for this change of attitude, mainly because farmers and the law have only recently demanded the improvements. Tractors such as the Fordson and the Ferguson achieved outstanding success because they provided what the majority of farmers wanted and were prepared to pay for: reliable, efficient machines with a minimum of frills or luxuries.

Various factors have combined to put greater emphasis on safety and comfort. One of these factors is the legislation which has forced manufacturers and their customers in many countries to fit safety cabs to new tractors. Evidence from Scandinavian countries and Britain is proving conclusively that these cabs reduce the numbers of fatal accidents occurring when tractors overturn. Similar legislation shows the need to reduce the noise level in tractor cabs to protect the hearing of drivers.

Evidence has been increasing since the 1950s of the problems which excessively noisy, badly designed cabs and tractor seats can cause over a long period of time. The hearing of tractor drivers tends to deteriorate more rapidly than the average for other occupations. A survey in East Germany has shown that the typically poor seating and vibration levels with which tractor drivers work tend to encourage spinal problems and stomach troubles. A survey in Britain showed that the rate of work is likely to be reduced when tractor drivers have to operate in conditions of excessive noise and vibration. This applies particularly to tractors with high horsepower engines. The driver, consciously or unconsciously, selects a lower engine speed than the optimum for the job, to reduce the discomfort involved. In areas where skilled drivers are in short supply, employers are increasingly prepared to pay for tractors with better cabs and seats.

Unfortunately there is no quick, easy answer to the problem of designing tractors which are really safe and comfortable to work in for long hours. With expensive, high horsepower models it is relatively easier to find the space for a roomy, properly insulated cab and to charge a price which will cover the extra cost which better design incurs. Smaller, cheaper tractors present greater problems. The cab is necessarily closer to the engine, and the cost of reducing vibration and insulating the cab from excessive noise, dust and exhaust fumes, becomes a greater proportion of the total cost of the tractor. There are special problems in designing a sprung suspension for tractors. This is partly because of cost and partly because of the loads which tractors may carry on the front or rear.

In spite of the problems, much progress has been made. Cabs are now being designed as part of the complete concept of new model tractors, instead of being afterthoughts hastily introduced to meet tightening legislation. The advantages of this approach to cab design are considerable, and some of the better cabs not only meet safety requirements and keep out the rain, but offer a high degree of comfort and convenience as well. Some cabs, especially on more expensive, high power tractors, have wide doorways with steps and grab rails, for easy access. Inside there should be a flat floor, preferably with a non-slip surface, and sufficient space and headroom for the driver to move about without being a contortionist. More fortunate drivers have tractor seats which adjust forwards and vertically, and which swivel and have springing adjustable for different weights. Steering wheels can sometimes be adjusted for height and tilt, and some other controls may be located in a console which moves as the seat is adjusted so that levers are always at hand. Plenty of window area gives good visibility all round, and rubber floor mats and acoustic panelling help to reduce noise.

Luxury extras on the most expensive tractors include multi-speed screen wipers, heaters, fresh air ventilation incorporating dust filters, cigarette lighters, radio and tape installations with volume, tone and balance controls, first-aid kits, anti-glare windscreens and interior lighting. These fittings, plus scientifically designed seats, and engine and transmission designs to minimise noise and vibrations, are still confined to a minority of tractors. But they are certain to become more general as tractor drivers begin to demand standards of comfort which truck drivers have enjoyed for years. The fact that legislation was required to start manufacturers designing for real safety is little credit to the tractor industry, but the standards which some manufacturers are now setting shows that the driver is no longer taken for granted. Now that safety and comfort have become sales points, competition among manufacturers to produce better conditions for the operator is a welcome development.

The tractor industry

During the early 1970s world production of tractors reached a balance of output between Russia and the East European countries, and the rest of the world. The available statistics are not always reliable, and are certainly not completely comparable, but they do indicate that the Communist world is drawing ahead of the capitalist world in the volume of tractors produced. In 1972 production in the rest of the world totalled about 700,000 tractors. This figure excludes garden tractors and also omits the relatively small number of tracklaying models produced. A very substantial proportion of this total is accounted for by the North American based companies, Allis-Chalmers, Case, John Deere, Ford, International Harvester, Massey-Ferguson and the White Motors group, Minneapolis-Moline, Oliver and Cockshutt.

Russia claimed to have produced 478,000 tractors in 1972, and production in other Communist bloc countries probably reached 250,000 or more. The Russian figure includes tracklaying versions, which are extremely important in Soviet agriculture, and it also includes some small horsepower models which would be classified as garden tractors in the United States. The Russians claim that their rate of production is increasing steadily, and their target for 1975 is 575,000 units a year. Very little is known about the volume of tractor production in China.

While Russian tractor production has been soaring, output from the United States manufacturers has tended to fall. Russian agriculture has been seriously short of tractor power and the effort to raise production is a reflection of this situation. The position is quite different in the United States and in many other countries with a high degree of farm mechanisation. Production of wheel and tracklaying agricultural tractors in the United States was less than 200,000 units a year during the early 1970s, but during the post war boom output topped 600,000 units in both 1949 and 1951. Production in 1970 was considerably less than in some of the better years of the 1920s.

American tractor production is limited by the fact that the domestic market is saturated. Farmers generally have as many tractors as they need, and new sales are made only when old tractors are scrapped. But the position is not simply a replacement market. Farmers are buying bigger, more powerful tractors, and often a single new, high horsepower tractor replaces two smaller tractors. In 1965 tractors of more than 100 h.p. accounted for only 2·3 per cent of new sales. By 1971 the same power range had increased its market share to twenty-five per cent of sales in the United States. From 1970 to 1971 sales of 140-h.p. plus tractors almost doubled to 2,549. Similar trends towards more powerful tractors are clearly evident in Canada, most European countries and other areas with mechanised farming. This tendency to use fewer, but more powerful tractors, is logical. As manpower becomes scarcer and more expensive, bigger tractors offer an obvious means of increasing productivity.

Russia has placed considerable emphasis on increasing tractor production in order to improve agricultural output and efficiency. By 1973 Soviet publicity was claiming that, under the control of a separate ministry, there were seventy-eight factories producing tractors with a range of wheel and crawler models from 7-h.p. horticultural machines up to the 300-h.p. DET–250. Most of the tractors produced are urgently required to boost Soviet farm mechanisation, but there is increasing emphasis on exports to earn foreign currency. The spearhead of the export sales effort is the Belarus range of wheel tractors produced in volume from the Minsk factory. According to the Soviet tractor exporting agency, these are operating in about sixty countries, with sales to developing countries and to the major tractor manufacturing areas of North America and Western Europe.

Tractors from Russia and from some other East European countries, such as Poland, Hungary, Rumania and Czechoslovakia, sell partly on the basis of low price, and partly on a reputation for ruggedness. The Belarus MTZ–50, for example, has earned respect in Britain for its performance in competitive ploughing events where efficiency is measured. In spite of their attractive price, tractors from Eastern Europe were making only a modest sales impact in countries such as Britain and Canada. Generally, these tractors have lacked the technical refinement of British or American models, and they are often less stylish in appearance. An exception to this generalisation is the Zetor range from Czechoslovakia which, during the early 1970s, achieved a good reputation in some European countries, including Britain, for its relatively advanced specification and also because it had a well-designed, roomy, comfortable cab with a low noise level.

It is likely that the average power of tractors on farms will continue to increase for many years, and at the other end of the horsepower scale, production of small tractors—30 h.p. or less—will continue to decline. The lower end of the power range is now such a small market that most American and British manufacturers have given up the production of small agricultural tractors, leaving a gap which Japanese companies are exploiting. There is still a place for the small tractor. Enormous numbers of 8Ns, Fergusons, Allis-Chalmers Bs and other popular tractors of less than 30 h.p. are still used on small farms. The ideal replacement for these ageing tractors is not always a second-hand big tractor. A light, economical, manoeuvrable tractor has obvious advantages in horticultural work, and also as a general-purpose tractor on large farms. A big acreage obviously demands the power of big tractors for tillage and harvesting jobs, but there are situations where the smaller tractor can operate more efficiently and more economically. These include the daily work of slurry scraping in large dairy units, transporting feed to livestock, and spreading fertiliser. Although the demand for small tractors is certainly declining, it is still big enough to generate an attractive volume of sales for Japanese tractors of 15 to 30 h.p. These have become a significant factor in the world market, competing against the Leyland 154 and the few other small models still produced by the established tractor companies in Europe and America. Mitsubishi and Kubota are names which are becoming increasingly familiar in America, supplementing rather than competing with the products of the American manufacturers. If the gap at the bottom of the market between garden tractors and small farm tractors is allowed to become too wide, the American tractor industry could have a Volkswagen situation on its hands one day.

7

TRACTORS IN THE FUTURE

Tractor development over the next ten years or so is reasonably predictable in a number of aspects. Engine power is likely to continue increasing and four-wheel drive will gain popularity. General standards of safety and comfort will increase on new models, and designers will probably pay more attention to styling. In other words tractors will be basically similar to those now in production, but bigger and better.

Longer term prediction about tractors towards the end of the century involves so many unknown factors that it becomes little more than rather complicated guesswork. Some of these unknown factors could mean fundamental changes to the economics, scale and methods of farming, resulting in changes in the design and function of tractors.

One of the question marks facing agriculture is the extent to which man-made substitutes for farm products will continue to gain wider acceptance. There are already substitutes for almost every livestock product marketed. Some of these manufactured products, including man-made fibres, margarine and imitation leathers, have already made massive commercial progress. Others, such as milk substitutes and meat analogues, are still facing flavour and texture problems. Demand for these alternative products could be stimulated sharply by excessive price increases for farm products, or by a reaction against foods involving slaughter.

A growing problem in countries with a mainly urban population is the increasing demand for land for housing, industry and recreation. This continuing trend is raising land values and encouraging research into alternative methods of food production which release land for other uses. One alternative is to 'farm' areas of the sea to produce fish or plankton.

Plants for food can be grown by hydroponic methods, with solutions containing the nutrients for plant growth replacing soil. This would be a factory process, with the crops produced throughout the year in a completely controlled environment.

The increasing cost of power appears to be a more immediate problem. Fuel costs have been relatively low because new reserves of oil have been discovered to keep pace with increasing consumption. Oil has been so plentiful that developed countries have been able to use it on a liberal scale. The United States, with six per cent of the world's population, has been accounting for almost one-third of the world's energy consumption.

Exploration to discover new reserves of oil is becoming increasingly costly because the more promising and more accessible areas have been covered. Oil producing countries are aware of the growing strength of their position. Some, already immensely wealthy from their oil revenues, are deliberately curbing production because oil they sell in the future will be worth so much more than oil sold now. Importing countries have shown their willingness so far to pay increasing prices for their oil, and there is little doubt that the price will continue to rise in the long term.

Faced with rising oil costs, and with the likelihood that reserves will be almost exhausted within about forty years, engineers are already investigating alternative power sources for vehicles. Before the end of the century we shall probably have tractor engines operating on fuels which do not originate from oil wells. During the Second World War it was shown that engines could operate on fuel derived from potatoes. It was calculated that two acres of potatoes could operate a 20-h.p. tractor in normal farm use for a year. Presumably this included the fuel used to grow and harvest the potatoes. In France, because of the war-time shortage of conventional fuel, some tractors were operated on the gas produced by the fermentation of manure. In 1941 Renault produced a 'gazogene' model, known as Model AFV-H, which had two large chambers for the fermentation gases mounted on each side of the tractor.

Powering tractors with methane produced from manure might be forced on farmers again eventually, and this might also be an answer to manure disposal problems. But there are other more promising alternatives awaiting evaluation.

Each year the world's engineering industry manufactures millions of internal combustion engines. Most of these work on the same basic principles which Nikolaus Otto first used a century ago when he developed the four-stroke cycle. Several attempts have been made to replace the four-stroke engine, particularly for providing power for vehicles, but so far the Otto type of engine has resisted every challenge.

The four-stroke engine has retained its supremacy because it is an extremely satisfactory power source for cars, trucks and tractors. Development and refinement have provided standards of reliability and economy which few of us question, and some modern cars operate with such smooth silence that it is easy to forget that the engine runs with a series of powerful explosions. Steam and electric power have both achieved modest, short-lived commercial success, and in the early 1950s there was great excitement about the possibilities of the gas turbine engine for cars and trucks. This, too, died down, leaving the four-stroke engine still completely dominant.

Recently the interest in alternative engines has been revived and the motor industry is investing heavily in research projects which could eventually put the four-stroke engine out of business. The main incentives for this research programme are the long-term need to end our dependence on dwindling oil supplies, and the short-term pressure to reduce the level of pollution caused by exhaust gases.

If the conventional four-stroke engine is superseded for cars and trucks, it would almost certainly be replaced for tractor power as well. Air pollution from tractors is not a significant environmental problem, but future legislation could be applied to tractors used for industrial applications. The cost of manufacturing four-stroke engines would increase if it was no longer virtually subsidised by the enormous volume of component production for the car and truck industries, and this would encourage tractor manufacturers to change from conventional engines.

Various rotary engines are under consideration for cars, at the time of writing, with the Wankel engine actually in commercial production. The outstanding advantages of the Wankel engine are claimed to be small size, light weight and smoothness of operation. These features appeal more to the motorist than to the farmer buying a tractor. The Wankel engine should also be cheaper than a four-stroke engine of comparable power, partly because of its lighter weight, and partly because the number of working parts is very much reduced. A 185-h.p. Wankel engine has about 150 moving parts compared with 350 to 400 in a V–8 of similar power. Critics of the Wankel unit point out that the potential cost reduction has

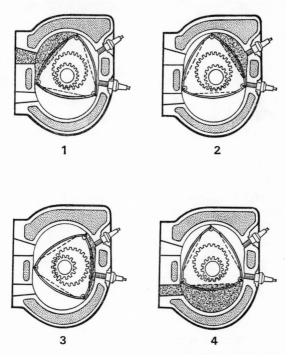

The Wankel rotary engine, operating cycle. 1, Induction, 2, Compression, 3, Ignition, 4, Exhaust.

yet to be achieved on a commercial scale, and that higher fuel consumption outweighs the engine's ability to operate satisfactorily on low octane fuels. A major question mark is the degree of pollution from the Wankel's exhaust, and there is some doubt as to whether this engine is likely to be any less offensive to the environmentalists than a conventional piston engine.

The Rover Motor Company announced the world's first gas turbine powered car in 1952, so earning themselves massive world-wide publicity. The company, now part of the British Leyland Motor Company, had worked in conjunction with Sir Frank Whittle on the commercial application of the jet engine, and had a head-start over their competitors. They announced a series of prototype cars during the 1950s and collaborated with B.R.M. to produce a gas turbine sports car which performed well during the 1965 Le Mans 24-hour race. Other car and truck manufacturers quickly joined Rover on the publicity bandwagon by announcing gas turbine prototypes, and there was widespread acceptance of the idea that turbine power for the family motorist was just a few years away. Chrysler in America actually produced a batch of fifty turbine-powered cars which were loaned out to motorists for a large-scale evaluation.

Since then there has been an ominous silence, indicating that adapting the turbine for volume production for millions of motorists was meeting problems, and only recently has there been a revival of interest. In theory, the gas turbine looks a more attractive proposition than anything else known. It should have a working life several times longer than a conventional piston engine—a mixed blessing to a motor industry geared to the relatively short operating life of the four-stroke engine. The turbine will operate on virtually any liquid which burns, engine maintenance would be greatly reduced, vibration is negligible and engine cooling is much simpler than with a conventional water cooled unit. Turbines also offer relatively quiet operation and less exhaust gas pollution.

Gas turbines offer special advantages for truck power, which could be attractive for agricultural tractors as well. They are particularly suitable for situations where sustained high power operation is needed, and can produce the very high horsepowers which heavy trucks and high-power tractors demand. Several of the world's leading truck manufacturers have been developing and testing turbine-powered vehicles, and some, including British Leyland and Ford, also have tractor manufacturing interests to which their turbine development could rapidly be applied.

International Harvester built and publicised a turbine powered tractor in 1961, which is displayed at the Smithsonian Institution in Washington D.C. This experimental model, the HT–340, was fitted with a Titan T62T 80-h.p. turbine, and the specification included hydrostatic transmission. International Harvester, with refreshing frankness, admitted that the

HT–340 was nowhere near the production stage, was noisy at idling speeds and used excessive quantities of fuel.

Commercial development of gas turbine units for vehicles and tractors has been delayed by problems of finding materials which will stand up to the high temperatures at which these engines operate, but without excessive cost. Manufacturers are naturally secretive about their progress, but, at the time of writing, it seems reasonable to guess that the gas turbine as a commercial proposition for large trucks is imminent, and turbine-powered tractors seem likely within ten years or less.

Electrically powered vehicles, which appeared briefly around the beginning of this century, have recently begun to attract interest again because of their quietness and absence of air pollution. Battery operated vehicles, which can be recharged overnight from a low-rate mains current, have some limited use for localised delivery work, and possibly for the so-called city car, which would be used for short distances only.

In 1959 Allis-Chalmers announced a fuel cell tractor, which was operated as a prototype and was later donated to the Smithsonian Institution. A fuel cell powered vehicle carries a suitable source of energy—Allis-Chalmers used a mixture of gases including mainly propane—which is fed into cells. The cells are similar to those of a battery, but the electricity, generated by a chemical reaction, is not stored. The Allis-Chalmers prototype had 1,008 cells, giving an electrical output of 15 kilowatts which powered a 20-h.p. D.C. motor. The tractor was claimed to produce 3,000 lb. drawbar effort under dynamometer tests, and was demonstrated in a number of operations, including ploughing.

Steam power enthusiasts have existed since before the development of Nikolaus Otto's first four-stroke engine. While steam power has disappeared from farms, and from railways and roads, the enthusiasts have been sustained by rumours, and occasional real evidence, that there would be a great revival one day. This idea is still discussed, and some research continues.

A future steamer, however, would be quite different from those of the past. Instead of a fire burning coal or straw, a future steam unit would have a bank of burners operating on liquid or gas fuel. These burners could be efficient enough to cause little air pollution, and the number operating would be directly related to the power required. The massive boiler and funnel of a steam ploughing engine would be replaced by a much smaller amount of water, or other liquid, such as freon, in a tube. Efficiency would be improved, and the weight of liquid carried reduced by returning steam from the pistons to a condenser where it would be cooled for repeated re-use, requiring infrequent topping up.

Steam engines of the future might be either piston or turbine designs. A steam turbine operates efficiently under a constant load, as with stationary work, or for operating a tractor p-t-o, but pistons are more suitable for traction, and have good torque characteristics at low speeds. A steam tractor might have both types of engine. Steam enthusiasts are not easily discouraged, but their hopes for a new generation of steam-powered vehicles do not, at present, appear well founded, in spite of the real progress which has been made. However, steam would be one possible means of using atomic energy if this power source is ever to be used directly for vehicles, including tractors. An atomic car or tractor would leave the factory powered for life, according to those who forecast the distant future, and would generate heat which would be converted to usable power by means of a steam engine. Before such a development becomes practical, however, someone must devise a safe method of disposing of the radio-active waste from the worn-out vehicles. A dealer's yard full of disintegrating atomic-powered tractors could be extremely dangerous.

Driverless tractors

The idea of a farm tractor that could be operated automatically without a driver has appealed to some engineers and economists for many years. Like many other 'new' ideas, the driverless tractor first appeared quite early in the history of tractor development.

One of the first attempts to automate field work was the 4-h.p. tractor plough developed in the early 1920s by a team working at the Iowa State College in America. Their machine was equipped with two single-furrow plough units attached to the tractor frame, one designed for right hand and the other for left hand working. The guidance and control systems were mechanical. A first furrow was ploughed conventionally with the driver steering, and this provided a line for the machine to follow automatically.

The unit was steered by a wheel in the previous furrow, and the direction of travel was controlled by reversing arms at each end of the tractor. When the machine reached the edge of a field, the impact of the reversing arm against a fence or other obstruction automatically reversed the direction of travel and also lifted one of the ploughs out of work and put the other plough into the working position. The machine would then travel back down the field until another obstacle tripped the reversing mechanism again. This form of mechanical control had obvious limitations and it appears to have got no further than the experimental stage.

Another relatively early American development appeared in 1931, when a Farmall 30 tractor equipped with remote control by radio signals was operated publicly. In theory at least, this is still considered to be a possibility for the future. Some attempts at forecasting long-term developments in farming suggest that one man may eventually operate a number of tractors from a central office. He would monitor their work by means of closed-circuit television and control their operation with radio signals.

Guidance systems based on signals from buried wires appear to be the most promising approach to tractor automation at the time of writing. This was demonstrated in Britain in 1957 at Reading University, and a prototype version of a similar system, developed by Peter Finn-Kelcey, was operated at the 1970 Royal Show using a Massey-Ferguson 165 tractor. In the same year the system was installed on an Essex fruit farm, probably the first use of driverless tractors on a commercial basis in the world. Finn-Kelcey's automatic control system has also been installed commercially in cider apple orchards in Britain. Other countries have also shown interest in using it in similar circumstances.

The wires carrying the signals are laid in a grid representing the path which the tractor must follow. They are buried below normal tillage depth and carry current from a 12-volt battery. The signal is detected by a sensing device mounted on the front of the tractor which operates a control box—the brain of the system. Signals from the grid can operate the tractor steering, clutch and brakes, change the forward speed and raise or lower implements on the three-point linkage. The signal can also be used to operate equipment, such as a sprayer, pulled by the tractor, and is capable of very precise control.

The equipment, as developed by Finn-Kelcey, fits on to a conventional tractor, and can easily be isolated when the tractor is required for manual operation. A number of safety devices are built into the system. If the tractor strays off course, runs into anything or develops a fault such as a flat tyre or low oil level, the engine cuts out automatically. If the guidance equipment fails, the tractor stops, and the safety cut-out can also be used to monitor the correct operation of equipment following the tractor. Costs can look attractive in relation to man-hours saved, especially for fruit growing and large-scale horticultural production where numerous tractor operations can be performed over a set course.

Situations which might be suitable for a driverless tractor include the care of grass on playing fields and recreation areas, and operating feed wagons in feedlots. For more general field operations, especially on a mixed farming system, automation of tractor work appears to have distinctly limited possibilities. It is difficult to believe that a driverless tractor would ever be really safe without constant human supervision. The news that a tractor was driving itself would attract most of the children for miles around, with the inevitable danger that they would try to get on it. Another difficulty is that a grid would be unlikely to cater for all of the patterns of tractor operation, particularly in a rotational farming system. A grid laid down for cultivating and planting potatoes would probably not suit haymaking work in the same field a year or so later. Some jobs, such as repeated passes with tined implements, should be carried out in varying directions.

To be a real match for a good human operator, the automatic guidance system must do more than simply steer a tractor around a planned course and switch off when something goes wrong. This is adequate, and makes good economic sense, in certain situations. But farming work generally involves thought, and a thoughtful tractor driver could not easily be replaced by buried wires and a control box. A good tractor driver will cure a squeak in a machine before it develops into a costly breakdown, check the wind direction before spreading liquid manure, and leave any damp areas of a hayfield until last when baling. Most of the farm tractors in the foreseeable future will have drivers at the controls.

8

PRESERVING THE PAST

The development of farm tractors has played a major role in revolutionising farming methods and improving the efficiency of food production. Tractor power has already made an enormous contribution to our living standards and offers a way of improving the lives of millions in developing countries.

Fortunately many people have had the foresight to preserve old tractors. Large numbers have been rescued from scrap heaps and breakers' yards, or from half-forgotten resting places in old barns or under hedgerows. Repaired, repainted and cared for, these tractors represent every phase of development, from the early experimental period of the nineteenth century and the heavyweights which first challenged steam on the prairies, through to more recent mass-produced models which show how power can be applied more effectively and more economically through the power-take-off, three-point linkage and rubber tyres. The history of the development of these machines, which have given us so much, is already well represented, and the record becomes more comprehensive each year as the number of tractors restored increases.

Some of the outstanding examples of tractor development are safely preserved for the future in museums. In London, the Science Museum, in a small but selective display, covers some of the outstanding developments. This includes one of the few surviving examples of an Ivel tractor, interesting because it was made in 1902, the year in which these tractors first reached the production stage. There is also a Fordson, one of the first batch produced for Britain's war effort, and the 1935 Ferguson Black Tractor made by Harry Ferguson as the prototype to demonstrate the system which almost every modern tractor now features.

The Smithsonian Institution in Washington, D.C. also has a 1918 Fordson. The large number of early examples of this model that have survived demonstrates their durability. The Smithsonian collection includes the Hart-Parr Number 3, built in 1903, which was effectively the prototype for the 17–30 series which established Hart-Parr in the commercial tractor business. There is also a Waterloo Boy Model N of 1918, and two recent prototype tractors, the Allis-Chalmers fuel cell model of 1959 and the International Harvester Gas Turbine tractor of 1961.

Henry Ford was fascinated by history, in spite of the statement 'history is bunk', which is attributed to him. He was particularly interested in the history of technology and made ample funds available for purchasing items for the Greenfield Village and Henry Ford Museum, which he started at Dearborn, Michigan. Tractor exhibits at the Museum include the 1907 prototype tractor which Ford assembled mainly from car components and, almost inevitably, a 1917 Fordson.

A type of museum which has become popular over the last twenty years or so is the collection of machinery and equipment illustrating the development of agriculture and rural life. These appeal to farming people with memories of tools they once used, and also to those from towns wishing to spend a few hours in a completely unfamiliar world. These museums are usually privately owned, and financed by admission fees. In Britain, at the time of writing, there are probably a dozen of this type and some, such as the Bicton Gardens Museum near Exmouth, Devonshire, and the Breamore House Museum, near Salisbury, Wiltshire, have excellent collections of tractors, dating mainly from the American influx of Titans, Fordsons, etc., during the First World War. In both these collections the tractors are well displayed and excellently restored.

British tractor enthusiasts have few opportunities to see the enormous tractors produced mainly in the United States for prairie farming conditions before and during the First World War. There are plenty of them in Canada, however, especially in the agricultural museums in the prairie provinces. The Manitoba Agricultural Museum at Austin listed some seventy tractor exhibits in 1973, including dozens of big ploughing tractors such as a 1912 30–60-h.p. Pioneer weighing 23,000 lb., and a 30–60-h.p. Rumely of 1910 alleged to weigh 28,000 lb. Ironically, this museum also has one of the few surviving British made tractors of this type, a 1914

Marshall. The Alberta museum is at Wetaskiwin, and is based on a private collection of agricultural items, again representing the era of giant tractors. The Western Development Museum at Saskatoon, Saskatchewan, features steam engines prominently, with a Reeves engine capable of pulling twenty furrows, in pride of place. But they also have a remarkable collection of more than 250 early tractors, claimed to be the best in North America. This collection includes a 60-90-h.p. Twin City tractor made in 1919, one of the biggest tractors ever built. It has a horizontal tubular cooling tank the size of a large steam traction engine boiler and driving wheels seven feet in diameter.

Tractor manufacturers vary considerably in the interest they display in preserving examples of their own history. John Deere have a Froelich replica of the 1894 model, and the Oliver Corporation have a 30-60-h.p. Hart-Parr Old Reliable of 1913 displayed at their Charles City headquarters. Some tractor companies have taken considerable trouble to research and document their history, and some are also prepared to give assistance to anyone requiring help in the restoration of obsolete models.

The main strength of interest in tractor history and preservation is among the thousands of enthusiasts in many countries who collect tractors and material, such as early catalogues, instruction books and photographs, relating to tractor history. Clubs and societies for people interested in preserving early tractors, often associated with steam engine and vintage farm machinery interests, already exist in Britain and in other countries, such as Canada and New Zealand, and their numbers are increasing. Members of these organisations provide the tractors, engines and other items for show exhibits, rallies and parades which demonstrate the history of farm mechanisation to a total audience of millions each year.

Restoring a rusting, derelict tractor to something approaching new condition is a job which requires hours of effort, skill and cash. The main reward is the satisfaction of achievement, but there can be a financial reward as well. Most old tractors are gaining value steadily, and this trend seems likely to continue as interest grows. The chances of discovering an early tractor are becoming rapidly smaller in Britain and in North America, although in Europe, where interest is not as well developed, the supply of old tractors is much greater. In America the more recent tractors which could be worth preserving include the Ford 9N. In 1973, at least, examples of this model in fair condition sold at auction for around $400, while specimens in poor condition fetch little more than $100. In Britain, a pre-war Fordson in a condition worth restoring can still be bought for less than £50 in some parts of the country. David Brown Cropmasters and Ferguson TE 20s, still in use on thousands of farms, are already finding their way into collections. Even if tractors of this period gain value only slowly, they can earn their keep by doing a few days' work when required.

1 Froelich, 1894.

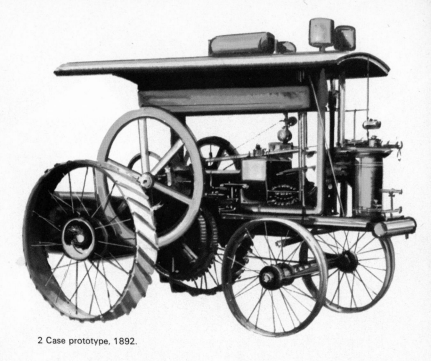

2 Case prototype, 1892.

3 Hornsby-Akroyd Patent Safety Oil Traction Engine, 1897.

4 Hart-Parr No. 1, 1902.

5 Hart-Parr No. 3, 1903.

6 Hart-Parr 30–60, Old Reliable.

7 Ford prototype, 1907.

8 Saunderson, 1910.

9 Ivel, 1902.

10 International Harvester Mogul, 1910.

11 Deutz Pfluglokomotive, 1907.

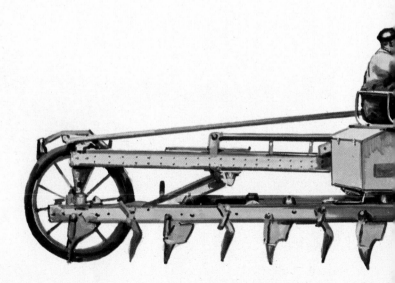

12 Hanomag, 1912.

13 Holt, 1910.

14 Pioneer, 1909.

15 Fowler, 1920.

16 Canadian, 1920.

17 International Harvester Titan, 1915.

18 International Harvester Titan, 1915.

19 Marshall, 1914.

20 International Harvester Mogul, 1915.

21 (*Top left*) Waterloo Boy Model N., 1918.

22 (*Above*) Parrett, 1916.

23 (*Left*) Bates Steel Mule C, 1915.

24 Munktells prototype, 1913.

25 International Harvester Mogul, 1914.

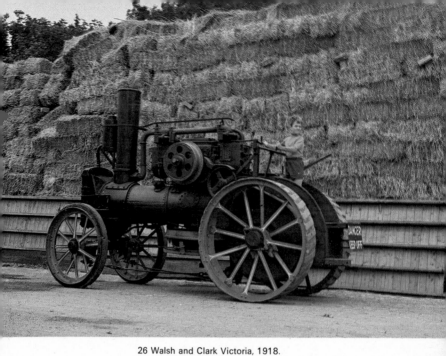

26 Walsh and Clark Victoria, 1918.

27 Wallis Cub Junior, 1916.

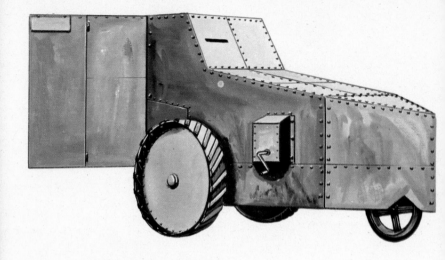

28 Ivel military prototype, 1910.

30 Daimler-Horse, 1917.

29 Killen-Strait military prototype, 1915.

31 Renault G.P., 1919.

32 Bryan, 1923.

33 Lanz Bulldog, 1921.

34 Somua, 1920.

35 Saunderson Universal, 1916.

36 Emerson-Brantingham, 1917.

37 Fordson, 1917.

38 Moline Universal, 1917.

39 Crawley, 1918.

40 Deutz MTH 222, 1924.

41 Cassani, 1927.

42 Aebi self-propelled mower, 1915.

43 Munktells, 1926.

44 Fendt self-propelled mower, 1928.

45 Tractor trials.

47 Fowler Gyrotiller, 1933.

46 Austral Auto Cultivator, 1920.

48 Glasgow, 1918.

49 International Junior, 1919.

50 Garner, 1919.

51 Austin, 1919.

52 Clayton, 1917.

53 International Farmall, 1925.

54 Lanz Bulldog, 1928.

55 Fiat, 1919.

56 Weeks Dungey, 1920.

57 John Deere Model D, 1923.

58 Rushton, 1929.

59 Oliver Hart-Parr, 1930.

60 John Deere GP, 1928.

61 Massey-Harris four-wheel drive, 1930.

62 International Harvester W 12, 1930.

63 International Harvester 10–20, 1928.

64 Minneapolis-Moline UTS, 1935.

65 Minneapolis-Moline RTS, 1939.

66 Ferguson Black Tractor, 1933.

67 Ferguson-Brown, 1936.

68 Ford 9N, 1939.

69 Ford 8N, 1947.

70 Tractor racing.

71 Marshall, 1936.

72 Minneapolis-Moline Orchard model, 1936.

73 Allis-Chalmers Universal, 1933.

74 Ford prototype, 1936.

75 Renault type AFV-H, 1941.

76 Oliver 80, 1944.

77 Deutz Bauernschlepper, 1936.

78 John Deere Model A, 1933–1947.

79 John Deere Model R, 1949.

80 John Deere L, 1939.

81 Allis-Chalmers B, 1948.

82 Farmall Model A, 1941.

83 David Brown 2D, 1956.

84 Allis-Chalmers G, 1949.

85 Case LA, 1942.

86 Fordson Rowcrop, 1942.

87 (*Above*) Fordson Standard, 1937.

88 (*Right*) John Deere M, 1947.

89 John Deere 720, 1956.

90 David Brown Cropmaster, 1951.

91 Ferguson TE, 1952.

92 Massey-Ferguson 35, 1958.

93 Fordson Major, 1946.

94 Fordson Major with half-track conversion, 1946.

95 Marshall, 1949.

96 Doe 3-D, 1964.

97 Leyland 154, 1971.

98 International Harvester Gas Turbine prototype, 1961.

99 Lely Hydro 90, 1970.

100 Dutra D4KB, 1969.

101 Fendt One-man System, 1965.

102 Valmet 702, 1969.

103 International Harvester 574, 1972.

104 County Forward Control, 1970.

105 David Brown 1212, 1972.

106 Fiat 605C Crawler tractors, 1970.

107 New Idea Uni-System, 1970.

108 New Idea Uni-System.

109 Aebi Transporter, 1972.

110 Aebi Transporter, 1972.

111 Massey-Ferguson 1200, 1973.

112 Mercedes-Benz M-B Trac, 1972.

113 Deutz Intrac, 1972.

114 Deutz Intrac, 1972, with rear mounted chemical tank.

115 Deutz Intrac, 1972, with front mounted implement.

116 Deutz Intrac, 1972, planting and spraying equipment.

117 Vantage, 1971.

118 Lely Supertrac, 1972.

1 Froelich, 1894. A replica of the Waterloo Gasoline Engine Company tractor, of which two were sold. The tractor was based on the Froelich prototype of 1892, with a single-cylinder vertical engine of 14 inch bore and stroke, developing 20 b.h.p. One forward gear and one reverse.

2 Case prototype, 1892. The first Case tractor, based on William Patterson's design. The 2-cylinder horizontal engine developed 20 b.h.p. Ignition and carburettor troubles discouraged commercial development.

3 Hornsby-Akroyd Patent Safety Oil Traction Engine, 1897. Manufactured by Ruston and Hornsby of Grantham, Lincolnshire, with a three-speed transmission and a pulley for stationary work, and powered by an 18-b.h.p. engine. The first tractor to be sold commercially in Britain.

4 Hart-Parr No. 1, 1902. Two-cylinder, four-stroke engine, 9 inch bore and 13 inch stroke, developing 30 b.h.p. The cooling system used oil and included the five expansion bulbs above the centre of the tractor.

5 Hart-Parr No. 3, 1903. Two-cylinder horizontal engine with make-and-break ignition and water injection to prevent 'knock'. The tractor weighed 7 tons and developed 30 b.h.p. It is now preserved in the Smithsonian Institution, Washington, D.C.

6 Hart-Parr 30–60, Old Reliable. This model was produced commercially for several years from 1907. Petrol/paraffin engine developing 60 b.h.p. at 300 r.p.m. The tractor weighed 10 tons, including the 1,000 lb. flywheel which was turned by hand to crank the engine. Photographed at the Manitoba Agricultural Museum, Austin, Manitoba.

7 Ford prototype, 1907. Henry Ford's first tractor, made up with wheels from a binder, the 4-cylinder petrol engine from a Model B car, and radiator, front axle and steering mechanism from a Model K car. The tractor developed 20 b.h.p., and is preserved at the Greenfield Village and Henry Ford Museum, Dearborn, Michigan.

8 Saunderson, 1910. Four-cylinder engine, of 6 inch bore and 8 inch stroke, starting on petrol and running on paraffin. The tractor weighed 4 tons 13 cwt., and had a 38-gallon fuel tank and 24-gallon capacity cooling system. Based on a photograph in the R.A.S.E. *Journal*, 1910.

9 Ivel, 1902. Two-cylinder horizontal engine, rated at 14 b.h.p., driving through one forward and one reverse gears. This was changed in later models for a 20-b.h.p. engine with two forward gears. This model, now displayed at the Science Museum, London, weighed 30 cwt. with a full tank of water for ballast and cooling.

10 International Harvester Mogul, 1910. The first model in the famous Mogul series, with a 2-cylinder horizontally opposed engine rated at 45 b.h.p. running at 345 r.p.m. Cooled by means of the 'water tower' at the front. A single forward speed—giving 2 m.p.h. maximum—and one reverse.

11 Deutz Pfluglokomotive, 1907. 40-b.h.p. ploughing tractor, with two ploughs attached for two-way working. The steering wheel and other controls were centrally located, with a driving seat on either side. The driver switched from one seat to the other to ensure that he faced forwards, as the tractor worked each way without turning. Based on a contemporary photograph provided by K.H.D., Cologne.

12 Hanomag, 1912. Probably the largest of the various tractor ploughs produced in Germany at about this time. The 4-cylinder engine had 15 litres capacity and developed 80 b.h.p., running on petrol.

13 Holt, 1910. 65-b.h.p. crawler tractor powered by a Holt 4-cylinder engine. Two forward speeds and one reverse. Weight, 21,300 lb. Based on photographs of the machine preserved at the Reynolds Museum, Wetaskiwin, Alberta.

14 Pioneer, 1909. 30–60-h.p. tractor made at Winona, Minnesota. Four-cylinder engine, horizontally opposed. Three forward gears and one reverse. Weight, 23,000 lb. Based on photographs of the tractor displayed at the Reynolds Museum, Wetaskiwin, Alberta.

15 Fowler, 1920. The version of the Fowler tractor, designed for cable ploughing, which took part in the 1920 R.A.S.E. tractor trials. The original version appeared in 1912, and was criticised for imitating the appearance of steam ploughing engines.

16 Canadian, 1920. Manufactured by the Alberta Foundry and Machine Company at Medicine Hat. Twin-cylinder horizontal engine, developing 14 h.p. at the drawbar, and with one forward gear ratio and one reverse. The main frame of the tractor is a 10 inch square beam of timber, and the wheel spokes are also made of wood. Based on photographs of the machine at the Reynolds Museum, Wetaskiwin, Alberta.

17 International Harvester Titan, 1915. 15–30-h.p. tractor, with a 4-cylinder paraffin engine equipped with high-tension magneto ignition. Transmission by spur gear and double chains. Two forward gear ratios, giving maximum speeds of 1·9 and 2·4 m.p.h.

18 International Harvester Titan, 1915. Two-cylinder engine with make-and-break ignition and hit-and-miss governor. This was the 30–60 model, developed from the 45-b.h.p. Titan of 1911. Weight 20,300 lb.

19 Marshall, 1914. 70-b.h.p. ploughing tractor, manufactured by Marshall of Gainsborough, Lincolnshire, and in this case, exported to Canada. Two forward gear ratios and one reverse, with the gear change mechanism located so that the driver had to dismount to operate it. Preserved at the Manitoba Agricultural Museum, Austin, Manitoba.

20 International Harvester Mogul, 1915. The 8–16 model, with single-cylinder engine operating at 400 r.p.m., with hopper cooling, and planetary gear transmission.

21 Waterloo Boy Model N, 1918. A descendant of the Froelich and immediate ancestor of the John Deere tractor range. Imported into Britain under the name Overtime. 25-b.h.p., 2-cylinder engine. This example is at the Smithsonian Institution.

22 Parrett, 1916. 12–25 h.p., with a 4-cylinder Buda engine operating at 1,000 r.p.m. Magneto ignition. Two forward gear ratios and one reverse. Manufactured by the Parrett Tractor Company of Chicago.

23 Bates Steel Mule C, 1915. 13–30-h.p. petrol/paraffin engine, of four cylinders. Driven by the single rear track, and with telescopic steering wheel shaft, so the driver could work from the seat of a trailed plough.

24 Munktells prototype, 1913. Eight-ton tractor manufactured by Munktells, with six-feet driving wheels. The original machine is preserved in working order by the manufacturer, now part of the Swedish Volvo organisation.

25 International Harvester Mogul, 1914. Twin-cylinder o.h.v. engine. Transmission by chain, with two forward speeds and a reverse. Weight approx. 90 cwt.

26 Walsh and Clark, Victoria, 1918. Made at Guiseley, Yorkshire, and designed for cable ploughing, with the cable drum mounted beneath the 'boiler' or fuel tank. Twin cylinder engine developing about 45 b.h.p.

27 Wallis Cub Junior, 1916. 13–25 h.p. developed at 650 r.p.m. This was a forerunner of the Massey-Harris tractor, and is considered to be the first production model with frameless construction.

28 Ivel military prototype, 1910. Fully armoured experimental version of Albone's tractor, developed for possible military use, but abandoned at an early stage.

29 Killen-Strait military prototype, 1915. Believed to be the first tracklaying armoured vehicle built. Imported from America as part of the British development programme which eventually produced the tank. The tractor, powered by a Waukesha petrol engine of 25–40 h.p., was fitted with the body shell of a Delaunay–Belleville armoured car.

30 Daimler-Horse, 1917. Two-wheel traction unit developed during the First World War to replace horses for moving war equipment. It weighed 3,750 lb., with a 4-cylinder engine developing 14·5 h.p. After the war these horse substitutes were adapted for farm work, leading to the Puch motor plough, and eventually to the Austrian Steyr-Daimler-Puch tractor range.

31 Renault G.P., 1919. Agricultural tractor developed directly from the famous Renault light tank. 30-h.p. petrol engine, with three forward gears and one reverse. This was further developed as the HI tracklaying model of 1923, with a wheeled version, the HO, introduced in 1927.

32 Bryan, 1923. An American attempt to revive steam power for farming. The power unit is a high-pressure tubular boiler.

33 Lanz Bulldog, 1921. Single-cylinder hotbulb engine of semi-diesel design, developing 12 h.p. This was the first small diesel-powered tractor to achieve substantial sales, and was the forerunner of a long line of Bulldog models using this engine type.

34 Somua, 1920. 35-b.h.p. engine. One of a range of powered rotary cultivators produced by this French company. Weight 2,450 kg., maximum travel speed 4·5 m.p.h.

35 Saunderson Universal, 1916. Twin-cylinder engine developing 25 h.p. at 750 r.p.m. Made at Bedford by one of the oldest of the British tractor manufacturers.

36 Emerson-Brantingham, 1917. Manufactured at Rockford, Illinois, and in this case exported to Britain to help satisfy demand for tractor power as the war ended. 26·4-h.p., 4-cylinder engine. The rear wheel track was narrower than the front, in order to avoid running in th e previous plough furrow.

37 Fordson, 1917. The first of thousands of Fordsons. 20 b.h.p. at 1,000 r.p.m. from a 4-cylinder engine. Frameless construction. This example is displayed at the Greenfield Village and Henry Ford Museum, Dearborn, Michigan.

38 Moline Universal, 1917, with Martin cultivator attached. Manufactured by the Moline Plow Company, Moline, Illinois to the design of the Universal Tractor Company of Columbus, Ohio. Twin-cylinder, o.h.v. engine mounted on a simple chassis designed so that implements could be attached centrally.

39 Crawley, 1918. 30-b.h.p., 4-cylinder engine on a frame designed to take a plough and other implements. A conversion kit was marketed to make the Crawley suitable for towing from the rear. Manufactured at Saffron Walden, Essex.

40 Deutz MTH 222, 1924. 14-h.p. semi-diesel engine, designed mainly for stationary work.

41 Cassani, 1927. Powered by a 40-b.h.p. diesel engine, the first tractor designed by Francesco Cassani, founder of the Italian Same company. Based on the tractor displayed at the Leonardo da Vinci Museum, Milan.

42 Aebi self-propelled mower, 1915. Manufactured in small numbers by a Swiss company specialising in mowing equipment. Powered variously by 3·5-h.p. New Way, 6·5-h.p. Felix or 10-h.p. EMW engines.

43 Munktells, 1926. Model 22HK, 4-cylinder vertical engine, frameless design. Manufactured at Eskilstuna, Sweden.

44 Fendt self-propelled mower, 1928. Power for propulsion and to operate the cutter-bar from a diesel engine. A prototype lightweight tractor plough, also diesel powered, was produced in the same year.

45 Tractor trials. Blackstone crawler in the foreground competing against a Clayton. Typical of the scene at many tractor trials held soon after the war in Britain. Based on contemporary photographs of the 1919 trials held near Lincoln.

46 Austral Auto Cultivator, 1920. The first rotary cultivator built by A. C. Howard, Moss Vale, Australia. A 60-b.h.p. Buda engine provided power for propulsion and to operate the 15 foot cultivator unit.

47 Fowler Gyrotiller, 1933. One of the later series, powered by a 137-b.h.p. diesel engine made by M.A.N. of Germany. The two powered rotor units at the rear give a tillage width of 10 ft., with a working depth of nearly 2 ft.

48 Glasgow, 1918. Four-cylinder engine driving all three wheels. 27 b.h.p. at 650 r.p.m. Weight 36 cwt. Cost new £450.

49 International Junior, 1919. 8–16 h.p. from a 4-cylinder engine. 30 cwt. Economical lightweight tractor made in America, but popular in Europe.

50 Garner, 1919. 30-b.h.p., 4-cylinder engine. Weight 30 cwt., three forward gear ratios and reverse. One of several British makes to appear in the tractor boom following the First World War.

51 Austin, 1919. 25-h.p., 4-cylinder, side-valve engine. Weight 30 cwt. Achieved brief commercial success in Britain, and later manufactured in France.

52 Clayton, 1917. 35-h.p. Dorman 4-cylinder engine, two forward gear ratios and reverse. One of the first British crawler tractors, and unusual in having a steering wheel.

53 International Farmall, 1925. 20-h.p., 4-cylinder engine. Outstandingly successful design for row-crop work, which helped International to meet the competition of the Fordson.

54 Lanz Bulldog, 1928. 24-h.p., single-cylinder diesel engine. One of the large number of Bulldog models produced by Lanz, with this type of engine. Equipped with a six-speed gearbox.

55 Fiat, 1919. Type 702, 30-h.p., 4-cylinder petrol engine. This was the first Fiat tractor, and approximately 2,000 of this model and the more powerful 703 were built before being replaced by the 700 in 1925. Cost new in Britain, £600.

56 Weeks Dungey, 1920. Another lightweight British tractor making a brief commercial appearance in the 1920s. Four-cylinder engine.

57   John Deere Model D, 1923. This model introduced the John Deere name into the tractor market. 15–27-h.p., twin-cylinder horizontal engine.

58   Rushton, 1929. Four-cylinder engine, 14–20 h.p. This tractor was built at Walthamstow, Essex, and was introduced against strong competition from the Fordson, which it closely resembled. Weight 3,950 lb. Cost new £209.

59   Oliver Hart-Parr, 1930. 18–28-h.p., 4-cylinder o.h.v. engine, $4\frac{1}{8}$ inch bore and $5\frac{1}{4}$ inch stroke, operating at 1,190 r.p.m. Offered with a choice of eight different lugs on the driving wheels.

60   John Deere GP, 1928. Claimed to be the first tractor offered with four power sources—drawbar, belt, p-t-o and powered implement lift. The arched front-axle design was to allow extra clearance for rowcrop work.

61   Massey-Harris four-wheel drive, 1930. One of the first attempts to put four-wheel drive into large-scale production. 24-h.p., 4-cylinder side-valve engine. Two forward ratios.

62   International Harvester W 12, 1930. 25 cwt. tractor rated at only 12 h.p. Four-cylinder engine, three forward ratios.

63   International Harvester 10–20, 1928. Equipped with p-t-o. The rubber tyres were not fitted until several years later.

64   Minneapolis-Moline UTS, 1935. Four-cylinder petrol/paraffin engine. One of the successful Universal series, which was available in 1941 with the option of an L.P. gas engine.

65   Minneapolis-Moline RTS, 1939. Four-cylinder paraffin engine, rated two-plough capacity.

66   Ferguson Black Tractor, 1933. The prototype built to demonstrate the Ferguson System. 18-h.p. Hercules engine, three-point linkage and automatic draft control. Displayed at the Science Museum, London.

67   Ferguson-Brown, 1936. The result of collaboration between Harry Ferguson and the David Brown Company of Huddersfield. 20-h.p. Coventry Climax engine in the first series, later replaced by a David Brown engine.

68   Ford 9N, 1939. The 'Ford tractor with Ferguson System'. 20-h.p., 4-cylinder engine, with three forward gear ratios, and full hydraulic control of the three-point linkage.

69   Ford 8N, 1947. Similar in appearance to the 9N, but with some improvements including an extra forward gear. The tractor which ended the Ford-Ferguson agreement.

70 Tractor racing. The publicity programme which helped to launch rubber tyres for tractors included speed trials and races. The races, between teams of specially prepared Allis-Chalmers U (Universal) tractors, became a popular spectacle at state fairs in 1933, with famous racing car drivers. Based on a contemporary photograph.

71 Marshall, 1936. Fitted with a continental-type single-cylinder diesel engine. 25 h.p. Weight 55 cwt.

72 Minneapolis-Moline Orchard model, 1936. A version of the J series, with bodywork designed for working between fruit trees and bushes, where conventional styling can damage branches.

73 Allis-Chalmers Universal, 1933. Originally introduced in 1929 with steel wheels. Low-pressure pneumatic tyres introduced in 1932. Four-cylinder engine, $4\frac{1}{2}$ inch bore and 5 inch stroke.

74 Ford prototype, 1936. An experimental tractor which followed Henry Ford's idea for making use of standardised parts. The front wheels were from the car production line, and the engine and radiator were standard Ford truck parts.

75 Renault type AFV-H, 1941. One of the famous gazogene tractors, modified for war-time fuel shortages, with two side-mounted containers for methane gas generated from decaying vegetable matter or manure.

76 Oliver 80, 1944. The 80 series, introduced in 1937, were claimed to have the first diesel engines fitted to a tractor—a claim which is difficult to substantiate. A Buda engine was used in early models, later replaced by an Oliver engine.

77 Deutz Bauernschlepper, 1936. This was a successful attempt to produce a tractor for the smaller European farms. 11 h.p., with rubber tyres and lighting available.

78 John Deere Model A, 1933–1947. Although dated in appearance when it was first introduced, this tractor remained in production, with some modification, for almost twenty years.

79 John Deere Model R, 1949. The first diesel engined tractor from John Deere, with a gasoline donkey engine to start the diesel.

80 John Deere L, 1939. 15-h.p., twin-cylinder engine, three forward gears and reverse. Weight 15 cwt.

81 Allis-Chalmers B, 1948. One of the most popular of the small, low-power tractors which were in fashion during the war years. 24 b.h.p., 1,400 r.p.m., 4-cylinder engine.

82 Farmall Model A, 1941. 15 h.p., three forward gear ratios and reverse.

83 David Brown 2D, 1956. Toolbar design with a 12-h.p., twin-cylinder air-cooled diesel engine. Implements attached to a central toolbar, which was raised by air-pressure operated by a compressor.

84 Allis-Chalmers G, 1949. Toolbar type tractor, with rear-mounted 15-h.p. four-cylinder engine, designed for market garden work.

85 Case LA, 1942. 56-h.p. paraffin engine. Four-cylinder side valve.

86 Fordson Rowcrop, 1942. 27 h.p., four-cylinder, petrol/paraffin.

87 Fordson Standard, 1937. A direct successor to the 1917 Fordson, but produced in England.

88 John Deere M, 1947. The first model produced from the then newly opened factory at Dubuque. Twin-cylinder, vertical engine.

89 John Deere 720, 1956. Rowcrop version of the new range of tractors introduced in 1956, with power steering and a padded seat to make the driver's life more pleasant.

90 David Brown Cropmaster, 1951. Available with paraffin or diesel engines, with the diesel version developing 34 b.h.p. at 1,800 r.p.m. Two-speed p-t-o.

91 Ferguson TE, 1952. The famous 'Fergie', produced as the TE series in England and the TO in America. Four-cylinder petrol/paraffin engine, but with a Perkins diesel engine available as an option, as on this version.

92 Massey-Ferguson 35, 1958. The first model introduced after the merger between Massey-Harris and Harry Ferguson. This tractor is working in a rice paddy in Malaysia.

93 Fordson Major, 1946. 27-h.p., four-cylinder engine. Either petrol/paraffin or diesel.

94 Fordson Major with half-track conversion, 1946. Half-track conversions were popular in Britain before 4-wheel drive tractors became generally available.

95 Marshall, 1949. Still powered by a single-cylinder diesel engine, which made the Marshall popular for stationary work.

96 Doe 3-D, 1964. Two Fordson Super Major engine units linked together in an articulated design, giving 100 b.h.p., 4-wheel drive and a high degree of manoeuvrability.

97 Leyland 154, 1971. Option of 25-h.p. diesel or 28-h.p. petrol engine. One of the few remaining non-Japanese tractors of this size.

98 International Harvester Gas Turbine prototype, 1961. Experimental 80-h.p.

gas turbine engine driving through a hydrostatic transmission. Displayed at the Smithsonian Institution, Washington, D.C.

99 Lely Hydro 90, 1970. 4·5 litre, 6-cylinder diesel engine, 87 b.h.p. at 2,600 r.p.m. Hydrostatic transmission.

100 Dutra D4KB, 1969. Successor to the Hungarian H.S.C.S. company. The Dutra is available in Britain with a choice of Csepol or Perkins engines.

101 Fendt One-man System, 1965. Toolbar type tractor designed to take a range of attachments which can be front, mid or rear-mounted. 45-h.p., three-cylinder diesel engine.

102 Valmet 702, 1969. 75-h.p., 4-cylinder diesel engine. This tractor, produced in Finland, was claimed to be one of the first to have a noise level of less than 90 db.

103 International Harvester 574, 1972. Hydrostatic transmission optional. Four-cylinder diesel engine.

104 County Forward Control, 1970. Ford 6-cylinder 102-b.h.p., diesel engine. The forward control arrangement is designed to give good visibility and load-carrying space behind the cab for carrying spray tanks or fertiliser.

105 David Brown 1212, 1972. 72-b.h.p. diesel engine. A choice of transmissions is available, including full synchromesh.

106 Fiat 605C Crawler tractors, 1970. The production line at the Turin factory. The 605C is 56 b.h.p., with six forward gears and one reverse, and is equipped with p-t-o and three-point linkage.

107 New Idea Uni-System, 1970. Based on the Minneapolis-Moline Uni tractor.

108 New Idea Uni-System. The basic power unit is matched by a range of special attachments, mainly for harvesting.

109 and **110** Aebi Transporter, 1972. A European approach to system mechanisation, designed especially for small farm units in Alpine areas. The transporter is available with a choice of single and twin-cylinder engines, and with a range of matched equipment. 6 forward gears, up to 16 h.p.

111 Massey-Ferguson 1200, 1973. The smallest of the three M-F four-wheel drive articulated tractors, with 120-h.p. diesel engine, and a cab designed as an integral part of the tractor.

112 Mercedes-Benz M-B Trac, 1972. Based on Unimog components, with four-wheel drive, and with front and rear p-t-o and implement attachment.

113, 114, 115 and **116**   Deutz Intrac, 1972. Like the M-B Trac, a German attempt to introduce new thinking to tractor design. Front and rear p-t-o and implement attachment, plus good visibility from the forward-control cab position. Choice of 90- or 51-b.h.p. engines.

117   Vantage, 1971. U.S. Steel Corporation idea of what future tractors may look like. Some similarities in layout to the Deutz Intrac.

118   Lely Supertrac, 1972. Prototype (November 1973) power unit, with a Perkins 178-h.p. engine, and ten forward speeds. Designed to take a matched range of implements.

FURTHER READING

Few books dealing with tractor history in detail were available in 1973, when this volume was being prepared. British tractors and some of the early tractor trials in Britain are covered by the Rev. Philip Wright in *Old Farm Tractors* published by David and Charles. Volume 7, numbers three and four of 'Industrial Archaeology', also published by David and Charles, included an authoritative two-part feature by Charles L. Cawood entitled *The History and Development of Farm Tractors*.

North American tractor history is described in *The Development of the Agricultural Tractor in the United States*, originally published by the U.S. Department of Agriculture in two volumes, and reprinted in 1956 by the American Society of Agricultural Engineers. *Power for Prairie Plows* by Grant McEwan, published by Prairie Press of Saskatchewan, provides a history of the introduction of tractors to Canadian prairie agriculture, and has an excellent section dealing with the Winnipeg Trials.

It would be possible to fill a small library with books dealing with Henry Ford I and the Ford companies' history. Not all of these are considered to be completely accurate, and few give more than a brief mention to the tractor interests of Ford. Allan Nevins and Frank Hill collaborated to write three books which give a comprehensive and detailed picture of Henry Ford and his industrial empire. These books, published by Scribner, include a considerable amount of information on the development of the Fordson, especially the second volume, *Ford: Expansion and Challenge, 1915–1933*.

There is also a good deal of information available about Harry Ferguson and his tractor activities. *Harry Ferguson, Inventor and Pioneer* by Colin

Fraser, published by John Murray, gives a particularly detailed and readable account of Ferguson and his contribution to history. *A Global Corporation* by E. P. Neufeld, published by the University of Toronto Press, also deals extensively with Ferguson. This book is also a source of detailed information about the development of the Massey-Harris company from the Second World War period through the formation of the Massey-Ferguson group.

This is not intended to be a comprehensive list of books dealing with tractor history, but covers a selection giving a comprehensive general outline of the subject.

INDEX